DNA DOUBLE HELIX

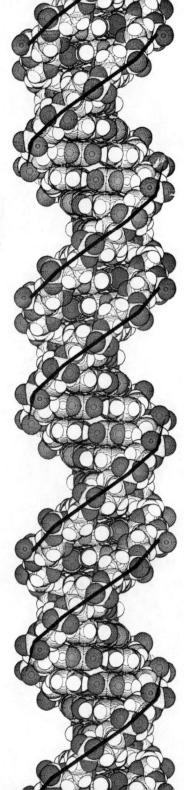

LIFE

ITSELF

ITS ORIGIN AND NATURE

Francis Crick

A TOUCHSTONE BOOK
Published by Simon and Schuster
NEW YORK

Grateful acknowledgment is made to the following for permission to reproduce diagrams.

From *Chemistry* by Linus Pauling and Peter Pauling. San Francisco: W. H. Freeman and Company, 1975.

From *Molecular Genetics: An Introductory Narrative,* 2nd ed., by Gunther S. Stent and Richard Calendar. San Francisco: W. H. Freeman and Company, 1978.

From *Biochemistry,* 2nd ed., by Lubert Stryer. San Francisco: W. H. Freeman and Company, 1981.

From *Science* magazine: "Left-Handed Double Helical DNA: Variations in the Backbone Conformation," Dr. Gary Guigley, Dept. of Biology, M.I.T., *Science,* Vol. 211, pp. 171–76, Cover, 9 January 1981.

From *Science* magazine: "Left-Handed Double Helical DNA: Variations in the Backbone Conformation," Wang, A. H. J., *Science,* Vol. 211, pp. 171–76, Cover, 9 January 1981.

From *Molecular Biology of the Gene,* 3rd ed., by James D. Watson. New York: W. A. Benjamin, Inc., 1976.

First Touchstone Edition, 1982

Published by Simon and Schuster
A Division of Gulf & Western Corporation
Simon & Schuster Building
Rockefeller Center
1230 Avenue of the Americas
New York, New York 10020

TOUCHSTONE and colophon are trademarks of Simon & Schuster

Designed by Eve Kirch

Manufactured in the United States of America

1 3 5 7 9 10 8 6 4 2
3 5 7 9 10 8 6 4 2 Pbk.

Library of Congress Cataloging in Publication Data

Crick, Francis, date.
Life itself.
Bibliography: p.

Includes index.
1. Life—Origin. 1. Title.
QH325.C84   577   81-9160
                                    AACR2

ISBN 0-671-25562-2
ISBN 0-671-25563-0 Pbk.

ACKNOWLEDGMENTS

THIS BOOK WAS WRITTEN after I moved to the Salk Institute in Southern California. I am grateful to the Kieckhefer Foundation for endowing a Research Chair for me and to the Ferkhauf and Noble Foundations for additional support. My especial thanks are due to the President of the Salk Institute, Dr. Frederic de Hoffmann, for providing me with an ideal environment for creative scientific work.

I doubt if I would have become involved with the problem of the origin of life but for my long friendship with Dr. Leslie Orgel. The idea of directed panspermia, which forms the skeleton of this book, originated in a joint paper we wrote together, but his influence has been more profound than that. His group at the Salk does experimental work on prebiotic chemistry and we discuss aspects of the problem almost every week. He also read an early draft and made careful comments. This draft was also read by Dr. Gustaf Arrhenius (the grandson of the Arrhenius who first suggested panspermia). As a result of his numerous comments several sections, especially those dealing with the primitive atmosphere of the earth, have been extensively rewritten, though he is not to be held responsible for the final result. Both Dr. Tom Jukes and my son Michael Crick have assisted in various ways.

As a rather inexperienced author I have been greatly helped by the staff of Simon and Schuster. Alice Mayhew's comments and suggestions have given the book a much better shape than it had originally. Her enthusiasm helped me to overcome my initial ditherings. She also suggested the title. Ann Godoff has been quietly efficient and very patient with confusions and postal delays. Nancy Schiffmann, the copy editor, managed, in the nicest way, to improve my English and to remove mistakes and ambiguities. My secretary at the Salk Institute, Betty Lars, struggled heroically with my almost unintelligible writing and especially with the many unfamiliar technical terms. To all of them, my thanks for their efforts.

For Odile

CONTENTS

12

Contents

PREFACE

So Where Are They?

THE ITALIAN PHYSICIST, Enrico Fermi, was a man with outstanding talents. His wife thought he was a genius and many scientists would agree with her. He was not only an exceptionally good theoretical physicist but also an experimentalist. It was Fermi and his friend, the Hungarian scientist Leo Szilard, who directed the design and construction of the first atomic pile, built in an unused squash court underneath a sports stadium in Chicago during World War II. In this unlikely environment the dangerous power of nuclear fission was harnessed for the first time on this planet.

Fermi, like most good scientists, had many interests outside his own particular field. He is credited with asking a famous question. There is a long preamble to Fermi's question, rather like a shaggy dog story. It goes something like this. The universe is vast, containing myriads of stars, many of them not unlike our sun. Our own galaxy has perhaps 10^{11}

stars* and there are at least 10^{10} galaxies and probably more. Many of these stars are likely to have planets circling around them. A fair fraction of these planets will have liquid water on their surface and a gaseous atmosphere made up of simple compounds of carbon, nitrogen, oxygen and hydrogen. The energy pouring down from the star—sunlight, in our case —onto the surface of the planet will cause the synthesis of numerous small organic compounds, thus turning the ocean into a thin, warm soup. These chemicals will eventually join onto each other and interact in intricate ways to produce a self-reproducing system, a primitive form of life. These simple living things will multiply, evolve by natural selection and become more complicated till eventually active, thinking creatures will emerge. Civilization, science and technology will follow and before long they will have mastered the entire environment of their planet. Then, yearning for fresh worlds to conquer, they will learn to travel to neighboring planets and then to planets on nearby stars, choosing for their colonization those with favorable environments. Eventually they should spread all over the galaxy, exploring it as they go. These highly exceptional and talented people could hardly overlook such a beautiful place as our earth, with its ample supply of water and organic compounds, its favorable temperature range and all its other advantages. "And so," Fermi would say, coming to his overwhelming question, "if all this has been happening they should have arrived here by now, *so where are they?*" It was Leo Szilard, a man with an impish sense of humor, who supplied the perfect reply to Fermi's rhetoric. "They are among us," he said, "but they call themselves Hungarians."

---

* This notation is so convenient that I shall use it throughout this book without further explanation. 10^{11} simply means a number consisting of a one followed by eleven zeros. That is, 100 billion. So a thousand is 10^3, a million 10^6, a billion (American) is 10^9, and so on.

Most people would accept the general trend of Fermi's argument. The difficulties arise when one tries to estimate the probability of each step, to put in numbers. There is no really hard evidence that other stars have planets, although it certainly seems likely that they do. If planets exist, at least a few will probably have a favorable environment for the production of a good soup—a mixture of simple organic compounds in water. It is the next step which is at present so mysterious: the formation from the soup of a primitive, chemical, self-reproducing system.

Even if this did happen we do not know how likely it is for the long process of evolution to culminate in a higher civilization, nor exactly how much time this might take, nor whether such creatures would really explore the universe nor how far they would succeed in traveling. All the events in Fermi's scenario may indeed be happening, but some steps may be very rare and some stages possibly rather slow. This would easily explain why, so far, we do not appear to have had visitors here from outer space.

As long ago as the latter part of the last century a rather different idea for the origin of life on earth was suggested by the Swedish physicist Arrhenius. He proposed that life did not succeed in starting here by itself but was seeded by microorganisms wafted in from space. These primitive spores, originating elsewhere, were supposed to be gently propelled by the pressure of the light falling on them. He called this idea *panspermia,* meaning "seeds everywhere." At the moment this idea is in disfavor because it is difficult to see how viable spores could have arrived here, after such a long journey in space, undamaged by radiation.

In this book I explore a variant of panspermia which Leslie Orgel and I suggested a few years ago. To avoid damage, the microorganisms are supposed to have traveled in the head of an unmanned spaceship sent to earth by a higher civilization which had developed elsewhere some billions of years ago. The spaceship was unmanned so that its range would be as great as possible. Life started here when these organisms were dropped into the primitive ocean and began to multiply. We called our idea *Directed Panspermia,* and published it quietly in *Icarus,* a space journal edited by Carl Sagan. It is not entirely new. J. B. S. Haldane had made a passing reference to it as early as 1954 and others have considered it since then, though not in as much detail as we did.

Whether Directed Panspermia should be considered genuine science or merely a rather unimaginative form of science fiction I discuss in Chapter 13. Most of the book is concerned with detailing the various steps in Fermi's argument. It sticks rather closely to the scientific knowledge we have today, flimsy though that often is. Rather than solve the problem of the origin of life on earth I want to sketch the background against which any solution must stand. And what a background it is! From the minuteness of atoms and molecules to the vast panorama of the entire universe; from events which take place in an infinitesimal fraction of a second to the entire time span of time itself, from the Big Bang to the present; from the intricate interplay of organic macromolecules to the endless complexities of higher civilizations and higher technology. It is one of the charms of this otherwise frustrating topic that to come to grips with it, one needs to know something about so many aspects of this astonishing universe in which we find ourselves.

LIFE
ITSELF

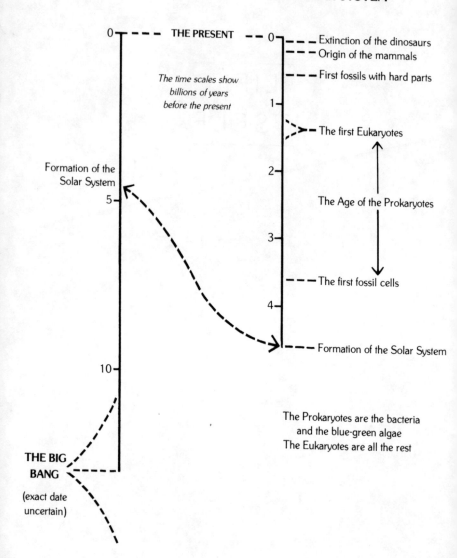

THE UNIVERSE

THE SOLAR SYSTEM

0 — — — THE PRESENT — — 0 — — Extinction of the dinosaurs
— — Origin of the mammals

*The time scales show
billions of years
before the present*

— — First fossils with hard parts

1

The first Eukaryotes

Formation of the
Solar System

5

2

The Age of the Prokaryotes

3

— — The first fossil cells

4

— — Formation of the Solar System

10

The Prokaryotes are the bacteria
and the blue-green algae
The Eukaryotes are all the rest

THE BIG
BANG

(exact date
uncertain)

ONE

Times and Distances, Large and Small

THERE IS ONE FACT about the origin of life which is reasonably certain. Whenever and wherever it happened, it started a very long time ago, so long ago that it is extremely difficult to form any realistic idea of such vast stretches of time. Our own personal experience extends back over tens of years, yet even for that limited period we are apt to forget precisely what the world was like when we were young. A hundred years ago the earth was also full of people, bustling about their business, eating and sleeping, walking and talking, making love and earning a living, each one steadily pursuing his own affairs, and yet (with very rare exceptions) not one of them is left alive today. Instead, a totally different set of persons inhabits the earth around us. The shortness of human life necessarily limits the span of direct personal recollection.

Human culture has given us the illusion that our memories go further back than that. Before writing was invented, the experience of earlier generations, embodied in stories, myths and moral precepts to guide behavior, was passed down verbally or, to a lesser extent, in pictures, carvings and statues. Writing has made more precise and more extensive the transmission of such information and in recent times photography has sharpened our images of the immediate past. Cinematography will give future generations a more direct and vivid impression of their forebears than we can now easily get from the written word. What a pity we don't have a talking picture of Cleopatra; it would not only reveal the true length of her nose but would make more explicit the essence of her charm.

We can, with an effort, project ourselves back to the time of Plato and Aristotle, and even beyond to Homer's Bronze Age heroes. We can learn something of the highly organized civilizations of Egypt, the Middle East, Central America and China and a little about other more primitive and scattered habitations. Even so, we have difficulty in contemplating steadily the march of history, from the beginnings of civilization to the present day, in such a way that we can truly experience the slow passage of time. Our minds are not built to deal comfortably with periods as long as hundreds or thousands of years.

Yet when we come to consider the origin of life, the time scales we must deal with make the whole span of human history seem but the blink of an eyelid. There is no simple way to adjust one's thinking to such vast stretches of time. The immensity of time passed is beyond our ready comprehension. One can only construct an impression of it from indirect and incomplete descriptions, much as a blind man laboriously builds up, by touch and sound, a picture of his immediate surroundings.

The customary way to provide a convenient framework for one's thoughts is to compare the age of the universe with the length of a single earthly day. Perhaps a better comparison, along the same lines, would be to equate the age of our earth with a single week. On such a scale the age of the universe, since the Big Bang, would be about two or three weeks. The oldest macroscopic fossils (those from the start of the Cambrian) would have been alive just one day ago. Modern man would have appeared in the last ten seconds and agriculture in the last one or two. Odysseus would have lived only half a second before the present time.

Even this comparison hardly makes the longer time scale comprehensible to us. Another alternative is to draw a linear map of time, with the different events marked on it. The problem here is to make the line long enough to show our own experience on a reasonable scale, and yet short enough for convenient reproduction and examination. For easy reference such a map has been printed at the beginning of this book. But perhaps the most vivid method is to compare time to the lines of print themselves. Let us make our entire book equal in length to the time from the start of the Cambrian to the present; that is, about 600 million years. Then each full page will represent roughly 3 million years, each line about ninety thousand years and each letter or small space about fifteen hundred years. The origin of the earth would be about seven books ago and the origin of the universe (which has been dated only approximately) ten or so books before that. Almost the whole of recorded human history would be covered by the last two or three letters of the book.

If you now turn back the pages of the book, slowly reading *one letter at a time*—remember, each letter is fifteen hundred years—then this may convey to you something of the immense stretches of time we shall have to consider. On this scale the span of your own life would be less than the width of a comma.

If life really started here we need hardly be concerned with the rest of the universe, but if it started elsewhere the magnitude of large distances must be squarely faced. Though it is difficult to convey a vivid and precise impression of the age of the universe, to grasp its size is almost beyond human comprehension, however we try to express it. The main stumbling block is the extreme emptiness of space; not merely the few atoms in between the stars but the immense distance from one star to another. The visible world close to us is cluttered with objects and our intuitive estimates of their distance depend mainly on various clues provided by their apparent size and their visual interrelationships. It is much more difficult to judge the distance of an unfamiliar object floating in the emptiness of the clear, blue sky. I once heard a Canadian radio interviewer say, when challenged, that he thought the moon "was about the size of a balloon," though admittedly this was before the days of space travel.

This is how two astronomers, Jastrow and Thompson, try to describe, by analogy, the size and the distance of objects in space:

> Let the sun be the size of an orange; on that scale the earth is a grain of sand circling in orbit around the sun at a distance of thirty feet; Jupiter, eleven times larger than the earth, is a cherry pit revolving at a distance of 200 feet or one city block from the sun. The galaxy on this scale is 100 billion oranges, each orange separated from its neighbors by an average distance of 1,000 miles.*

The difficulty with an analogy of this type is that it is almost impossible for us to estimate distances in empty

---

* More information on Jastrow and Thompson's book can be found in the Further Reading section at the end of the book.

space. A comparison with a city block is misleading, because we too easily visualize the buildings in it, and in doing so lose the idea of emptiness. If you try to imagine an orange floating even a mile up in the sky you will find that its distance seems to become indefinite. An "orange" a thousand miles away would be too small to see unless it were incandescent.

Another possible method is to convert distances to time. Pretend you are on a spaceship which is traveling faster than any present-day spaceship. For various reasons, which will become clear later, let us take its speed to be one-hundredth the velocity of light; that is, about 1,800 miles per second. At this speed one could go from New York to Europe in about three seconds (Concorde takes roughly three hours), so we are certainly traveling fairly fast by everyday standards. It would take us two minutes to reach the moon and fifteen hours to reach the sun. To go right across the solar system from one side to the other—let us take this distance rather arbitrarily as the diameter of the orbit of Neptune—would take us almost three and a half weeks. The main point to grasp is that this journey is not like a very long train journey, rather longer than the distance from Moscow to Vladivostok and back. Such a trip would probably be monotonous enough, even though the landscape were constantly flowing past the train window. While going across the solar system, there would be nothing at all just outside the window of the spaceship. Very slowly, day after day, the sun would change in size and position. As we traveled farther away from it, its apparent diameter would decrease, till near the orbit of Neptune it would look "little bigger than a pin's head," as I have previously described it, assuming that its apparent size, as viewed from the earth, corresponds roughly to that of a silver dollar. In spite of traveling so fast—remember that at this

speed we could travel from any spot to any other on the
earth's surface in less than seven seconds—this journey would
be tedious in the extreme. Our main impression would be of
the almost total emptiness of space. At this distance a planet
would appear to be little more than an occasional speck in
this vast wilderness.

This feeling of an immense three-dimensional emptiness is
bad enough while we are focusing on the solar system. (Almost all of the scale models of the solar system one sees in
museums are grossly misleading. The sun and the planets are
almost always shown as far too big by comparison with the
distances between them.) It is when we try to go farther
afield that the enormity of space really hits us. To reach the
nearest star—actually a group of three stars fairly close together—would take our spaceship 430 years and the chances
are we would pass nothing significant on the way there. A
whole lifetime of one hundred years, traveling at this very
high speed, would take us less than a quarter of the way
there. We would be constantly traveling from emptiness to
emptiness with nothing but a few gas molecules and an
occasional tiny speck of dust to show that we were not always
in the same place. Very, very slowly a few of the nearest stars
would change their positions slightly, while the sun itself
would fade imperceptibly until it was just another star in the
brilliant panorama of stars visible on all sides of the spaceship. Long though it would seem, this journey to the nearest
star is, by astronomical standards, a very short one. To cross
our own galaxy from side to side would take no less than ten
million years. Such distances are beyond anything we can
conceive except in the most abstract way. And yet, on a
cosmic scale, the distance across the galaxy is hardly any
distance at all. Admittedly it is only about twenty times as
far to Andromeda, the nearest large galaxy, but to reach the
limits of space visible to us in our giant telescopes we would

have to travel more than a thousand times farther than that. To me it is remarkable that this astonishing discovery, the vastness and the emptiness of space, has not attracted the imaginative attention of poets and religious thinkers. People are happy to contemplate the limitless powers of God—a doubtful proposition at best—but quite unwilling to meditate creatively on the size of this extraordinary universe in which, through no virtue of their own, they find themselves. Naïvely one might have thought that both poets and priests would be so utterly astonished by these scientific revelations that they would be working with a white-hot fury to try to embody them in the foundation of our culture. The psalmist who said, "When I consider Thy heavens, the work of Thy fingers, the moon and the stars, which Thou hast ordained; what is man, that Thou art mindful of him? . . ." was at least trying, within the limitations of his beliefs, to express his wonder at the universe visible to the naked eye and the pettiness of man by comparison. And yet *his* universe was a small, almost cozy affair compared to the one modern science has revealed to us. It is almost as if the utter insignificance of the earth and the thin film of its biosphere has totally paralyzed the imagination, as if it were too dreadful to contemplate and therefore best ignored.

I shall not discuss here how these very large distances are estimated. The distance of the main objects in the solar system can now be obtained very accurately by a combination of the theory of solar mechanics and radar ranging, the distances of the nearest stars by the way their relative positions change slightly when viewed from the different positions of the earth in its yearly orbit around the sun. After that the arguments are more technical and less precise. But that the distances are the sort of size astronomers estimate there is not the slightest doubt.

So far we have been considering very large magnitudes.

Fortunately, when we turn to very small distances and times things are not quite so bad. We need to know the size of atoms—the size and contents of the tiny nucleus within each atom will concern us less—compared to everyday things. This we can manage in two relatively small hops. Let us start with a millimeter. This distance (about a twenty-fifth of an inch) is easy for us to see with the naked eye. One-thousandth part of this is called a micron. A bacteria cell is about two microns long. The wavelength of visible light (which limits what we can see in a high-powered light microscope) is about half a micron long.

We now go down by another factor of a thousand to reach a length known as a nanometer. The typical distance between adjacent atoms bonded strongly together in an organic compound lies between a tenth and a fifth of this. Under the best conditions we can see distances of a nanometer, or a little less, using an electron microscope, provided the specimen can be suitably prepared. Moreover, it is possible to exhibit pictures of a whole series of natural objects at every scale between a small group of atoms and a flea, so that with a little practice we can feel one scale merging into another. By contrast with the emptiness of space, the living world is crammed with detail at every level. The ease with which we can go from one scale to another should not blind us to the fact that the numbers of objects within a *volume* can be uncomfortably large. For example, a drop of water contains rather more than a thousand billion billion water molecules.

The short time we shall be concerned with will rarely be less than a picosecond, that is, one-millionth of a millionth of a second, though very much shorter times occur in nuclear reactions and in studies of subatomic particles. This minute interval is the sort of time scale on which molecules are vibrating, but looked at another way, it does not seem so outlandish. Consider the velocity of sound. In air this is relatively slow—little faster than most jet planes—being about a thousand feet per second. If a flash of lightning is only a mile away, it will take a full five seconds for its sound to reach us. This velocity is, incidentally, approximately the same as the average speed of the molecules of gas in the air, in between their collisions with each other. The speed of sound in most solids is usually a little faster.

Now we ask, how long will it take a sound wave to pass over a small molecule? A simple calculation shows this time to be in the picosecond range. This is just what one would expect, since this is about the time scale on which the atoms of the molecule are vibrating against one another. What is important is that this is, roughly speaking, the pulse rate *underlying* chemical reactions. An enzyme—an organic catalyst—can react a thousand or more times a second. This may appear fast to us but this rate is really rather slow on the time scale of atomic vibration.

Unfortunately, it is not so easy to convey the time scales in between a second and a picosecond, though a physical chemist can learn to feel at home over this fairly large range. Fortunately, we shall not be concerned directly with these very short times, though we shall see their effects indirectly. Most chemical reactions are really very rare events. The molecules usually move around intermittently and barge against one another many times before a rare lucky encounter allows them to hit each other strongly enough and in the correct direction to surmount their protective barriers and produce a chemical reaction. It is only because there are usually so many molecules in one small volume, all doing this at the same time, that the rate of a chemical reaction appears to proceed quite smoothly. The chance variations are smoothed out by the large numbers involved.

When we stand back and review once again these very different scales—the minute size of an atom and the almost unimaginable size of the universe; the pulse rate of chemical reaction compared to the deserts of vast eternity since the Big Bang—we see that in all these instances our intuitions, based on our experience of everyday life, are likely to be highly misleading. By themselves, large numbers mean very little to us. There is only one way to overcome this handicap, so natural to our human condition. We must calculate and recalculate, even though only approximately, to check and recheck our initial impressions until slowly, with time and constant application, the real world, the world of the immensely small and the immensely great, becomes as familiar to us as the simple cradle of our common earthly experience.

TWO

The Cosmic Pageant

NOW THAT WE have become familiar with the magnitudes involved, both large and small, for both space and time, we must sketch what we know of the origin of the universe, together with the formation of the galaxies and the stars and finally of the planets which make up our solar system, so that we can outline the conditions under which life originated, either on earth or elsewhere in the cosmos.

If the origin of life is difficult to approach because it happened so long ago, it might be thought that the origin of the universe, which must have happened appreciably earlier, would be even more inaccessible. This is not entirely true, because the interactions needed to start a living system are a small intricate subset of many other possible interactions in a very heterogeneous environment, whereas during the earlier stages of the Big Bang everything was so intimately mixed together that it was the broad outlines of the reactions which in large part dominated the process. It is thus easier to come to grips with them.

Almost all recent discussions of the origin of the universe are based on the Big Bang theory. This postulates that at the first stage we can usefully think about, the entire substance of the universe may have occupied only a rather small volume at an immensely high temperature. This primeval fireball was expanding very rapidly, cooling as it did so. Steven Weinberg has written an excellent book outlining for the general reader the sort of reactions which are likely to have taken place in the first three minutes.*

The picture is built up from our present-day knowledge of the fundamental particles of matter and radiation, together with a rather small number of experimental facts, such as the cosmic radiation background which now pervades all space —the faint whisper of creation just audible in radio telescopes. Such an imaginative synthesis is necessarily not entirely secure. Weinberg confesses to an occasional feeling of unreality in writing about it. The other important observable facts needed to construct the theory are the expansion of the universe, shown by the famous red shift, and the enormous excess in the present universe of particles of electromagnetic radiation (photons) compared to particles of matter (baryons) —the ratio is about 10^9 (a billion) to one—plus the relative scarcity of the heavier elements. Even in the present universe ninety-nine percent of the atoms are accounted for by the two lightest ones, hydrogen and helium, the former being the more common. From all these facts theoretical physicists can deduce that after the first one-hundredth of a second (which is even more uncertain) the fireball was an intricate mixture of radiation and matter, interacting together rapidly and strongly at an immensely high temperature—about 10^{11} degrees—and expanding extremely fast. The temperature was far too high to allow atoms to exist, and even too hot for

* Weinberg, Steven, *The First Three Minutes*. New York: Basic Books, Inc., 1977.

complex nuclei (the dense centers of atoms) to hold together. As the fireball expanded, it cooled, passing in quick succession through several stages in which, because the temperature in each stage was lower than before, certain processes occurred less frequently and others became more common. Eventually, after about three minutes, the temperature was a mere 10^9 degrees, so that certain very light nuclei, such as those of tritium and helium, could now form without being broken apart. After half an hour or so the temperature had fallen to 3×10^8 (300 million) degrees—only twenty times hotter than the interior of the sun—and the synthesis of new nuclei stopped. For a further million years or so the universe went on expanding and cooling until the nuclei could capture electrons to form stable atoms. Matter could then start to condense into galaxies and stars.

Because of this enormous cosmic explosion the universe has been expanding ever since. Whether it will continue to expand indefinitely or whether it will eventually slow down till it stops and falls back on itself depends upon exactly how massive it is. Just as a stone thrown high into the air will fall back to earth unless it is thrown so fast that it can escape altogether, so the universe will go on expanding unless its mass is so big that in the end gravity will halt the expansion and reverse it. If this is so, at some time very far in the future the universe will collapse on itself in another catastrophic event. It used to be thought that the estimated density of the universe was too small to allow this—the critical density corresponds to about three hydrogen atoms in every liter of space. It is now suspected that those little neutral particles, the neutrinos, which pervade all space and which previously were thought, like light, to be weightless, may perhaps have a finite but very small mass. If so, there may be enough of them to stop the universe from expanding forever.

Perhaps the most important conclusion, from our limited

point of view, is that in the early stages of the universe, in spite of the very high density and temperature, only the very lightest elements were formed in any appreciable quantities. As a result, with the exception of hydrogen, all the elements vital to life, in particular carbon, nitrogen, oxygen and phosphorus, had yet to be made. This deduction is confirmed by spectroscopic observations, which show that the oldest stars have much less of these elements than the younger ones.

After the first million years the details of the picture become somewhat clouded. Exactly how the growing fireball, which is presumed to have been spatially rather uniform, expanded even further to produce the great heterogeneous clumps of matter we see as galaxies, and exactly how the various types of star were formed—these questions have not yet been answered in detail, though we can glimpse some of the processes in outline.

Whereas gravity played little part in the earlier stages of the universe, it now started to assume a more dominant role. In a broad way we can see that because of gravity, matter is likely to form into clumps which will attract other clumps, till eventually larger and larger aggregates are produced. The impacts involved in this accretion and condensation will raise the local temperature till the mass becomes so hot that it is luminous. Eventually the larger lumps of matter will reach such a high temperature that nuclear reactions will start—a star will have been formed.

From then on, the heat produced by nuclear fusion will prevent the star from collapsing on itself, since if this starts to happen the star will heat up, the nuclear reactions will go faster and the resulting increase of pressure will make the star expand a little to correct the incipient collapse. This mechanism acts as a regulator which permits the star to "burn" smoothly for many millions or even billions of years.

In the long run the star must run out of nuclear fuel. Calculations show that large stars burn up very fast, medium-sized stars (like the sun) more slowly and small stars very slowly indeed. A star ten times as massive as the sun runs through its fuel a hundred times more quickly. What happens as the nuclear fuel begins to run out is quite complicated and depends very much upon how massive the star is. The process of fusion may produce elements such as carbon and nitrogen from hydrogen and helium. The star may then try to use these heavier elements as fuel, producing even heavier ones, but eventually there comes a stage where there are no elements left whose transmutation can provide it with sufficient energy. At this point the all-embracing force of gravity, which has been kept in check by the heat generated by the nuclear processes, gets the upper hand. The star will collapse on itself. Exactly how this happens depends once again on the size of the star and the nature of the components. The smaller stars will probably end up as white dwarfs and very, very slowly fade from sight. For the larger stars, the collapse may be so rapid that the star literally explodes, spewing as much as half of itself into space and scattering matter at high speed in all directions. Many of the elements heavier than iron (which are not very abundant) are produced during the actual explosion itself.

Such a catastrophic explosion is called a supernova. For a matter of days the star shines exceedingly brightly. When this happened to a star in our own galaxy in 1604 it caused a sensation. We can still observe the remnants of an earlier supernova seen by Chinese astronomers in 1054. This great cloud of luminous gas, which we call the Crab Nebula, is still expanding rapidly, and we can even see the remnant of the star, now a pulsar (a rotating neutron star), at its center.

It is explosions like these which were the main source of

most of the elements in your body (hydrogen excepted). It gives one a strange feeling to realize that many of the atoms of which we ourselves are made were not formed at the beginning of things but had to be cooked up inside a star and scattered into space.

How, then, are planets formed? This is considered in a little more detail in Chapter 8. Here we shall only sketch the background. As we peer with our telescopes at the complexities of our own galaxy we can see that much of it is obscured by great clouds of gas and dust, some very diffuse, some less so, but all very tenuous by earthly standards. The particles of dust, about the same size as the particles in cigarette smoke, are probably made of tiny bits of iron, rock, ice and carbon compounds, mixed together. Rather surprisingly, over fifty types of small organic molecules have been discovered floating in these gas clouds, especially in the denser ones (where there is little ultraviolet light to damage them), though in mass they amount, in total, to only about one part in a million. These are chemically reactive molecules such as hydrogen cyanide (HCN) and formaldehyde ($HCHO$). Exactly what part this vast amount of very dilute molecules, scattered in space, played in the origin of life is uncertain, but their direct role is unlikely to have been a major one. The small molecules which form the basis of life (see Chapter 3 and also Chapter 5)—the amino acids, the sugars, the bases, etc.—have not yet been detected there, though some of them would be fairly easily synthesized from the ones that do occur in space. There is some speculation as to what reactions may have occurred in comets and other small bodies in the solar system.

It is believed that our sun and its attendant planets were formed by the condensation, due to gravity, of a slowly spinning cloud of this general sort. Exactly how this happened is still a matter of controversy. Roughly speaking, as the cloud collapsed its rate of rotation increased (to conserve angular momentum) so that it spun out into a disk. The center of this disk eventually became the sun while the remaining wisps of matter condensed to form the planets and the asteroids. The process is considered more fully in Chapter 8.

Much of this cloud must have consisted of hydrogen and helium, since these are the elements most abundant in the sun, but a planet like the earth is too near the sun and also not massive enough to retain such light elements by the pull of its own relatively feeble gravitational field, so they were presumably lost into space. (The large outer planets still have much of them.) The earth, with its inner core of iron and the solid skin of lighter elements near its surface, was built from the accumulated ashes of earlier stars. The biosphere in which we live is a frail veneer of matter on the surface of a rather small planet of a rather average star.

The most important point to emerge from this very brief sketch is that life as we know it could not possibly have arisen shortly after the Big Bang because the elements needed to construct it did not then exist. A period of some one or two billion years, possibly more, was required before enough large stars had run through their life cycle and exploded to provide the atoms needed for organic life. These had then to be swept up to form new stars and planets from the debris. Unfortunately, we do not know exactly how easy a process this is, so that we cannot be confident, on theoretical grounds, just how many stars are likely to have planets revolving around them, though as we shall see in Chapter 8 there is some indirect evidence on this point.

Let us now briefly recapitulate the sizes and times we are interested in. The diameter of the solar system is about a

1/1500 of a light-year. The nearest star is 4.3 light-years away. There are about a hundred stars within twenty light-years. Our own galaxy is a slowly spinning irregular disk of stars, dust and gas, about 100,000 light-years across, containing perhaps 10^{11} stars. The nearest large galaxy is Andromeda, somewhat bigger than ours. It is about two million light-years away, with very, very little in between (neutrinos and photons aside), though there are a few smaller galaxies in the general neighborhood. Beyond that the universe extends in all directions to a distance of at least three billion light-years, and contains a total of perhaps 10^{11} galaxies of various sorts and sizes.

The age of the earth and the rest of the solar system is about 4½ billion years. The time which has elapsed since the Big Bang is known with less precision but probably lies between seven and fifteen billion years. There were effectively none of the heavier elements shortly after the Big Bang but an appreciable supply of them was available a billion or so years later.

THREE

The Uniformity of Biochemistry

THE PROBLEM of the origin of life is, at bottom, a problem in organic chemistry—the chemistry of carbon compounds —but organic chemistry within an unusual framework. Living things, as we shall see, are specified in detail at the level of atoms and molecules, with incredible delicacy and precision. At the beginning it must have been molecules that evolved to form the first living system. Because life started on earth such a long time ago—perhaps as much as four billion years ago—it is very difficult for us to discover what the first living things were like. All living things on earth, without exception, are based on organic chemistry, and such chemicals are usually not stable over very long periods of time at the range of temperatures which exist on the earth's surface. The constant buffeting of thermal motion over hundreds of millions of years eventually disrupts the strong chemical bonds which hold the atoms of an organic molecule firmly together over shorter periods; over our own lifetime, for example. For this reason it is almost impossible to find "molecular fossils" from these very early times.

Minerals can be much more stable, at least on a somewhat coarser scale, mainly because their atoms use strong bonds to form regular three-dimensional structures. The failure of a single bond will not disturb the shape of the mineral too much. Fossils are seen in abundance in rocks laid down a little over half a billion years ago, at a time when organisms had evolved sufficiently to develop hard parts. Such fossils are not usually made of the original material of those organisms but consist of mineral deposits which have infiltrated them and taken up their shape. The shape of the soft parts is usually lost, though occasionally traces like wormholes are preserved—footprints on the rocks of time.

Are there any fossils much earlier than this? Careful microscopic examination of very early rocks has shown them to contain small structures which look like the fossilized remnants of very simple organisms, rather similar to some of the unicellular organisms on the earth today. This makes good sense. In the process of evolution we would expect creatures with many cells to develop from earlier ones having only single cells. Although there is still some controversy about the details, the earliest organisms of this type have been dated to about $2\frac{1}{2}$ to $3\frac{1}{2}$ billion years ago. The age of the earth is about $4\frac{1}{2}$ billion years. After the turmoil of its initial formation had subsided there was a period of about a billion years during which life could have evolved from the complex chemistry of the earth's surface, especially in its oceans, lakes and pools. Of that period we have no fossil record at all, because no preserved parts of the sedimentary rocks from that time have yet been found.

There are only two ways for us to approach this problem. We can try to simulate those early conditions in the laboratory. Since life is probably a happy accident which, even in the extended laboratory of the planet's surface, is likely to have taken many millions of years to occur, it is not too surprising that such research has not yet got very far, though some progress has been made. In addition we can look carefully at all living things which exist today. Because they are all descended from some of the first simple organisms it might be hoped that they still bear within them some traces of the earliest living things.

At first sight such a hope seems absurd. What could possibly unite the lily and the giraffe? What could a man share with the bacteria in his intestines? A cynic might wonder whether, since all living things eat or are eaten, this at least suggests they have something in common. Remarkably, this turns out to be correct. The unity of biochemistry is far greater and more detailed than was supposed even as little as a hundred years ago. The immense variety of nature—man, animals, plants, microorganisms, even viruses—is built, at the chemical level, on a common ground plan. It is the fantastic elaboration of this ground plan, evolved by natural selection over countless generations, which makes it difficult for us, in our everyday life, to penetrate beneath the outward form and perceive the unity within. In spite of our differences we all use a single chemical language, or, more precisely, as we shall see, *two* such languages, intimately related to each other.

To understand the unity of biochemistry we must first grasp in a very general way what chemical reactions go on within an organism. A living cell can be thought of as a fairly complex, well-organized chemical factory which takes one set of organic molecules—its food—breaks them down, if necessary, into smaller units and then reassorts and recombines these smaller units, often in several discrete steps, to make many other small molecules, some of which it excretes and some of which it uses for further synthesis. In particular, it strings special sets of these small molecules together into long chains, usually unbranched, to make the vital macromolecules of the cell, the three great families of giant molecules: the nucleic acids, the proteins and the polysaccharides.

The first level of organization we must consider is the lowest of all—that at which atoms are bound together to form small molecules. Now, a single atom is a fairly symmetrical object. Its shape is approximately spherical and if we look at it in a mirror it appears exactly the same, just as a billiard ball would. More intricate structures can have a "handedness"—our own hands are a good example. If we look at a right hand in a mirror we see a left hand, and vice versa. We can oppose our two hands, as in prayer, but this is as if we held a mirror between them. There is no way in which we can exactly superimpose one on the other, even in our imagination.

Some simple organic molecules, such as alcohol, have no "hand"; they are identical to their mirror images, as indeed a cup is. But this is not true of most organic molecules. The sugar on the breakfast table, if looked at in a mirror, becomes a significantly different assembly of atoms. This difference does not matter for *all* types of chemical reaction. If we heated such a molecule and could watch the molecular vibrations increase until one of the bonds broke, we would see that, had we imagined the mirror image of this process, the relative movements of all the atoms would have been identical. The basic reactions of chemistry are symmetrical under reflection to a very high degree of approximation. The difference in the hand only becomes important when two molecules have to fit together. We can see this in the manufacture of a glove. All the components of a glove—the fabric, the sewing thread, even the buttons—are, individually, mirror-symmetric, but they can be put together in two similar but different ways, to make either a righthanded glove or a left-handed one. Obviously we need two sorts because we have two kinds of hands—a good lefthanded glove will not fit properly onto a right hand.

The simplest form of asymmetrical molecule of this type arises when a single carbon atom is joined by single bonds to four other *different* atoms, or groups of atoms. This is because the four bonds of the carbon atom do not all lie in the same plane but are spaced out equally in all three dimensions, pointing approximately toward the corners of a regular tetrahedron.

Thus, organic molecules—molecules containing carbon atoms—often have a hand, even though they may be small, but we still have to realize why this matters in a cell. The

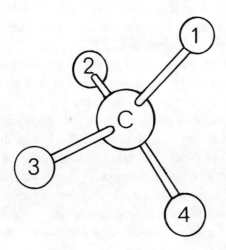

The distribution in space of the four bonds around a single carbon atom.

basic reason is that a biochemical molecule does not exist in
isolation. It reacts with other molecules. Almost every biochemical reaction is speeded up by its own special catalyst. A
small molecule, to react in this way, has to fit snugly onto
the catalyst's surface, and since the small molecule has a
hand, the catalyst must also have one. As in the case of a
glove, the reaction will not work properly if we try to fit a
lefthanded molecule into the cavity appropriate for a righthanded one.

Imagine you could watch this minute chemical factory
working and could see all the numerous reactions going on,
with molecules diffusing rapidly from one place to another,
fitting onto the various catalytic molecules, breaking, re-
forming, regrouping and reacting in many different ways.
Now imagine you were watching a factory which was the
exact mirror image of the first one. Everything would proceed
exactly as before, since the laws of chemistry are the same in
a mirror. Trouble would arise only if you tried to combine
the two, using some components from one system mixed
with others from the mirror world.

We can thus see why, in a single organism, the handedness
of the many asymmetrical molecules, large and small, must
be concordant. Moreover, it is an experimental fact that the
asymmetrical molecules on one side of your body have exactly
the same hand as those on the other side. But could we not
have two distinct types of organisms, one the mirror image
of the other, at least as far as its components are concerned?
This is what is never found. There are not two separate
kingdoms in nature, one having molecules of one hand and
the other their mirror images. Glucose has the same hand
everywhere. More significantly, the small molecules that are
strung together to make proteins—the amino acids—are all
L-amino acids (their mirror images are called D-amino acids:

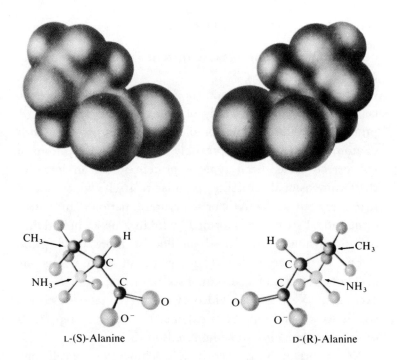

L-(S)-Alanine   D-(R)-Alanine

The two forms of the amino acid alanine. Each is the mirror image of the other. The upper figures use space-filling models; the lower over ball-and-spoke models. The letters indicate the atoms. The form of alanine found in proteins is L-alanine, the one on the left.

L = *L*evo, D = *D*extro) and the sugars in the nucleic acids are also all of one hand. The first great unifying principle of biochemistry is that the key molecules have the same hand in all organisms.

There are many other biochemical features which are astonishingly alike in all cells. The actual metabolic pathways —the precise ways in which one small molecule is converted into another—are often remarkably similar, though not always identical. So are some of the structural features, but the uniformity is even more striking at the deepest levels of organization; striking because there it is both arbitrary and complete.

Much of the structure and the metabolic machinery of the cell are based on one family of molecules, the proteins. A protein molecule is a macromolecule, running to thousands of atoms. Each protein is precisely made, with every atom in its correct place. Each type of protein forms an intricate three-dimensional structure, peculiar to itself, which allows it to carry out its catalytic or structural function. This three-dimensional structure is formed by folding up an underlying one-dimensional one, based on one or more polypeptide chains, as they are called. The sequence of atoms along this backbone consists of a pattern of six atoms, repeated over and over again. Variety is provided by the very small side-chains which stick out from the backbone, one at every repeat. A typical backbone has some hundreds of them.

Not surprisingly, the synthetic machinery of the cell constructs these polypeptide chains by joining together, end to end, a particular set of small molecules, the amino acids. These are all alike at one end—the part which will form the repeating backbone—but different at the other end, the part which forms the small side-chains. What is surprising is that there are just twenty kinds of them used to make proteins, and this set of twenty is exactly the same throughout nature. Yet other kinds of amino acids exist and several of them can be found within a cell. Nevertheless, only this particular set of twenty is used for proteins.

A protein is like a paragraph written in a twenty-letter language, the exact nature of the protein being determined by the exact order of the letters. With one trivial exception, this script never varies. Animals, plants, microorganisms and viruses all use the same set of twenty letters although, as far as we can tell, other similar letters could easily have been employed, just as other symbols could have been used to

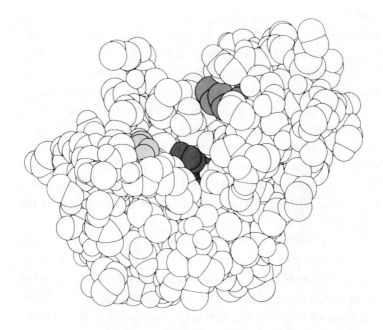

An atomic model of a small protein, the enzyme ribonuclease S. The shaded atoms form part of the active site of the enzyme. The protein would normally be entirely surrounded by water molecules.

construct our own alphabet. Some of these chemical letters are obvious choices, since they are small and easily available. Others are less obvious. If every printed text in the world used exactly the same arbitrary set of letters (which, as we know, is far from the case), we would reasonably conclude that the fully developed script had probably originated in one particular place and been passed on by constant copying. It is difficult not to come to the same conclusion for the amino acids. The set of twenty is so universal that its choice would appear to date back to very near the beginning of all living things.

Nature employs a second, very different chemical language

which is also fairly uniform. The genetic information for any organism is carried in one of the two closely related families of giant chain molecules, the nucleic acids, DNA and RNA, described in more detail in Chapter 5. Each molecule has an immensely long backbone with a regular, repeating structure. Again, a side-group is attached at regular intervals but in this case there are only four types; the genetic language has only four letters. A typical small virus, such as the polio virus, is about five thousand letters long. The genetic message in a bacterial cell usually has a few million letters; man's has several billion, packed in the center of each of our many cells.

One of the major biological discoveries of the sixties was the unraveling of the genetic code, the small dictionary (similar in principle to the Morse Code) which relates the four-letter language of the genetic material to the twenty-letter language of protein, the executive language. It is described in detail in the Appendix.

To translate the genetic message on a particular stretch of nucleic acid, the sequence of the side-groups is read off by the biochemical machinery in groups of three, starting from some fixed point. Since the nucleic acid language has just four distinct letters, there are sixty-four possible triplets ($4 \times 4 \times 4$). Sixty-one of these, codons as they are called, stand for one amino acid or another. The other three triplets stand for "end chain." (The signal for "begin chain" is a little complicated.)

The exact nature of the genetic code is as important for biology as Mendeleev's Periodic Table of the Elements is for chemistry, but there is an important difference. The Periodic Table would be the same everywhere in the universe. The genetic code appears rather arbitrary, or at least partly so. Many attempts have been made to deduce the relationship between the two languages from chemical principles, but so far none have been successful. The code has a few regular features, but these might be due to chance.

Even if there existed an entirely separate form of life elsewhere, also based on nucleic acids and protein, I can see no good reason why the genetic code should be exactly the same there as it is here. (The Morse Code, incidentally, is not completely arbitrary. The commonest letters, like *e* and *t,* are allocated the shortest number of dots or dashes.) If this appearance of arbitrariness in the genetic code is sustained, we can only conclude, once again, that all life on earth arose from one very primitive population which first used it to control the flow of chemical information from the nucleic acid language to the protein language.

Thus, all living things use the same four-letter language to carry genetic information. All use the same twenty-letter language to construct their proteins, the machine tools of the living cell. All use the same chemical dictionary to translate from one language to the other. Such an astonishing degree of uniformity was hardly suspected as little as forty years ago, when I was an undergraduate. I find it a curious symptom of our times that those who derive deep satisfaction from brooding on their unity with nature are often quite ignorant of the very unity they are attempting to contemplate. Perhaps in California there already exists a church in which the genetic code is read out every Sunday morning, though I doubt whether anyone would find such a bare recital very inspiring.

We see, then, that one way to approach the origin of life is to try to imagine how this remarkable uniformity first arose. Almost all modern theories and experimental work on life's origin take as their starting point the synthesis of either nucleic acid or protein or both. How could the primitive earth (if indeed life first started on earth) have produced the first relevant macromolecules? We have seen that these chain molecules are made by joining together small subunits end to end. How could the small molecules have been synthesized under early, prebiotic conditions? And how could we decide, even if we could have watched the whole operation in atomic detail, when the system first deserved to be called "living"? To come to grips with this problem we must examine next just what attributes we would expect *any* living system to have.

FOUR

The General Nature of Life

IT IS NOT EASY to give a compact definition of either "life" or "living." Certainly by "living" I do not necessarily mean thinking or feeling, since, to a biologist, plants are certainly alive and few people (apart from a few credulous individuals without scientific training) believe that plants think and feel as we and other animals do. Bacteria—and how little they must feel, even though they can "smell" food molecules and swim toward them—must certainly be considered alive. Viruses are more difficult. With them we come near to the borderline between the living and the nonliving. Perhaps the best way to approach the problem is to describe what we know of the basic processes of life, stripping away the skins of the onion until there is little or nothing left, and then to generalize what we have discovered.

When we do this we cannot help being struck by the very high degree of *organized complexity* we find at every level, and especially at the molecular level, since we have every reason to believe that structures easily visible to the naked eye, as well as those seen only with a microscope, are all built up from the intricate interactions of their molecular components. How complicated are these macromolecules and exactly how are they made?

The most remarkable example of molecular architecture found in living organisms is undoubtedly the protein family. Even a relatively simple protein may have as many as two thousand atoms, forming a fairly precise three-dimensional structure, with every atom in its particular place, except when disturbed by the constant jostling produced by thermal motion. Moreover, this intricate three-dimensional shape is essential for its function. If the molecule, in solution in water, is heated, in most cases the increased temperature will first loosen and then break the weak bonds holding the underlying chain in its correct fold so that the structure becomes jumbled and disorganized. No longer will it have the correct cavities, with the appropriate chemical groups, on its surface, and so it will no longer be able to fulfill its original function. If other protein molecules, also in this disorganized state, are in the solution, they may all stick together and coagulate, so that even when the solution is cooled the tangled mass cannot unravel itself. Boil an egg and the thick suspension of proteins becomes hopelessly mixed together, making it mechanically firm where before it was soft and runny.

At first sight it would seem a very difficult task to make an exact copy of the intact three-dimensional structure of a protein in its well-organized native fold. One could conceive making a molecular cast of the surface, as one might for a piece of sculpture, but how would one copy the inside of the molecule? Nature has solved this difficulty with a neat trick. The polypeptide chain is synthesized as an extended, rather one-dimensional structure and then *folds itself up*. The folding process is directed by the precise pattern of the side-chains, which interact together, and with the backbone, using multiple, weak interactions. The molecule explores the constant opportunities offered by thermal movement until, by trial and error, the best fold is discovered. The different parts of the molecule then slip neatly together, fitting so well that further thermal motion leaves the molecule relatively undisturbed.

To produce this miracle of molecular construction all the cell need do is to string together the amino acids (which make up the polypeptide chain) *in the correct order*. This is a complicated biochemical process, a molecular assembly line, using instructions in the form of a nucleic acid tape (the so-called messenger RNA) which will be described in outline in Chapter 5. Here we need only ask, how many possible proteins are there? If a particular amino acid sequence was selected by chance, how rare an event would that be?

This is an easy exercise in combinatorials. Suppose the chain is about two hundred amino acids long; this is, if anything, rather less than the average length of proteins of all types. Since we have just twenty possibilities at each place, the number of possibilities is twenty multiplied by itself some two hundred times. This is conveniently written 20^{200} and is approximately equal to 10^{260}, that is, a one followed by 260 zeros!

This number is quite beyond our everyday comprehension. For comparison, consider the number of fundamental particles (atoms, speaking loosely) in the entire visible universe, not just in our own galaxy with its 10^{11} stars, but in all the billions of galaxies, out to the limits of observable space. This number, which is estimated to be 10^{80}, is quite paltry by comparison to 10^{260}. Moreover, we have only considered a polypeptide chain of a rather modest length. Had we considered longer ones as well, the figure would have been even more immense. It is possible to show that ever since life started on earth, the number of different polypeptide chains which could have been synthesized during all this long time is only a minute fraction of the number of imaginable ones. The great majority of sequences can never have been synthesized at all, at any time.

These calculations take account only of the amino acid sequence. They do not allow for the fact that many sequences would probably not fold up satisfactorily into a stable, compact shape. What fraction of all possible sequences would do this is not known, though it is surmised to be fairly small.

A loose analogy may make this clearer. Consider a paragraph written in English. This is made from a set of about thirty symbols (the letters and punctuation marks, ignoring capitals). A typical paragraph has about as many letters as a typical protein has amino acids. Thus, a similar calculation to the one above would show that the number of different letter-sequences is correspondingly vast. There is, in fact, a vanishingly small hope of even a billion monkeys, on a billion typewriters, ever typing correctly even one sonnet of Shakespeare's during the present lifetime of the universe. Much of what was typed would be completely nonsensical. If we ask what *fraction* of possible paragraphs would have some sort of meaning, we find that this would also be minute. Nevertheless, the number of *meaningful* paragraphs is very great, even if we have no easy way of estimating this number. In the same way, the number of possible distinct, compact, stable proteins must be very large.

What we have discovered is that even at this very basic level there are complex structures which occur in many identical copies—that is, which have organized complexity—and which cannot have arisen by pure chance. Life, from this point of view, is an infinitely rare event, and yet we see it teeming all around us. How can such rare things be so common?

Stripped of its many fascinating complexities, the basic mechanism is very simple. It was suggested by both Darwin and Wallace, each of whom conceived the idea after reading Malthus. Living organisms must necessarily compete, for food, for mates and for living space, especially with other members of their own species. They must avoid predators and other dangers. For all these various reasons, some will leave more offspring than others, and it is the genetic characteristics of such preferred replicators which will be passed on preferentially to succeeding generations. In more technical terms, if a gene confers increased "fitness" on its possessor, then such a gene is more likely to be found in the gene pool of the next generation. This is the essence of natural selection. At first sight it seems almost a tautology; however, it is not the words that matter but the underlying mechanisms. Can we say, in very abstract terms, what they must be?

The first obvious requirement is for replication, and rather precise replication at that. We need to carry a considerable amount of information as instructions to form the complexity which characterizes life, and unless this information is copied with reasonable accuracy the mechanism will decay under the accumulated weight of errors. Perfect accuracy, on the other hand, is not a requirement. Indeed, all copies should not be exactly the same. Many of the copying errors will be a handicap but a few are likely to be an improvement, allowing the gene to function more effectively. We need these for natural selection to operate on. Thus, we need mutations, as these genetic errors are called, but not too many of them. In practice the error rate needed is exceptionally low, so low, in fact, that the cell usually has to take special precautions to correct most of the mistakes, leaving only a few to produce the variety needed if a species is to continue to hold its own and to evolve.

It is important to notice that the mutations must themselves be copied by the replicating mechanism. There is no use for mistakes which cannot be copied, for these would merely foul up the system. Such mistakes must be eliminated in some way. Confronted with such a chemical error, the copying system may ignore it and put in one of the standard letters at random. For natural selection to operate, it does not matter all that much what mistake is made as long as the end result is an alteration which can be copied faithfully in the succeeding generations.

Replication and mutation are the two essential requirements. It has been implied that a gene can be more or less "fit." The most minimal advantage it can have is that it can be directly copied more quickly or more often than its relations. Usually it achieves this end in less direct ways. It may direct the production of a messenger RNA which codes for a protein which has some special and desirable property, so that the organism that possesses it has an advantage in the struggle to produce more and better offspring. In technical terms, an improved gene will usually change not merely the genotype (the collection of genes in an organism) but also its phenotype (loosely speaking, the properties it exhibits to the world). This will usually be based on the properties or abundance of one or more proteins, since proteins control most of the chemical activities of the body, whereas nucleic acid, especially DNA, does very little except replicate and code for proteins and certain structural RNA molecules.

There is one final general requirement. We must avoid

"cross-feeding." In general, we do not want a rival organism to benefit from the product of our genes. We want those products to help only our own genes. This means that we must keep a gene and its products together in some way. At the lowest levels this is conveniently done by keeping the genes and most of their products in the same bag. This bag is called a cell and is surrounded by a very thin semipermeable membrane which prevents most of the molecules inside the cell from leaving it, unless there are good reasons why one is needed outside. Special gates and pumps exist in the membrane to get food and other molecules into the cell from outside or to let out waste products and other selected molecules.

This outlines the main demands on the informational system needed for life, but from them flow more immediate and mundane requirements. Since we need to make copies of some of the molecules we must have an adequate supply of raw material. Except in very special cases, these chemicals will need to be transformed into other, related chemicals. In modern cells each such step is usually catalyzed by a particular protein—an enzyme—specific for that reaction only. At the origin of life the raw material must have been mainly in a form ready for immediate use, since at that time there can have been few if any specific catalysts to make the primitive soup more palatable.

To carry out organic synthesis a supply of energy is needed, and this must be *available* energy. The technical term for this is *free energy,* which does not merely imply that you are getting it for nothing. (The term has a rather precise thermodynamic meaning.) The system is thus not in equilibrium, in the narrow sense of the term, though it may be in dynamic equilibrium. A very loose analogy would be to contrast a rather still pond, whose equilibrium is static, with a running river, which keeps on flowing steadily in much the same sort of way. A living system resembles the river. Material and free energy flow into it, while waste products and heat flow out. In technical terms, it is an open system. Only in this way can it continue to maintain the synthesis needed for repeated chemical replication.

These, then, are the basic requirements for life. The system must be able to replicate directly both its own instructions and indirectly any machinery needed to execute them. The replication of the genetic material must be fairly exact, but mutations—mistakes which can be faithfully copied— must occur at a rather low rate. A gene and its "product" must be kept reasonably close together. The system will be an open one and must have a supply of raw material and, in some way or another, a supply of free energy.

Stated in these broad terms the requirements do not seem too demanding, though, as we shall see, they are rather difficult to fulfill, starting from scratch. What is not quite so apparent is the marvelous capacity of such a system to improve itself. A copying process with a few rare errors—what could this possibly lead to?

The first thing to grasp is the continuing nature of the process. To achieve anything striking, the system must effectively go on forever. But this implies that we are doubling the number of copies every "generation." Most people are familiar with the idea that this rapidly leads to unmanageable numbers. The traditional story concerns the king or sultan who wished to reward one of his subjects and asked him what he would like. The man (it is unclear whether he was cunning or naïve, wise or foolish—tyrants do not usually like to be made to look silly) is alleged to have made what at first sight seems a very modest request. Pointing at the chessboard, he asked for a single grain of wheat for the first square, two for the second, four for the third, eight for the next and so on, doubling each time. This may not seem unreasonable until one recalls that a chessboard has sixty-four squares. A little simple algebra shows that the number of grains of wheat needed is one less than 2^{64}. This is a little more than 10^{19}, corresponding to a weight of about 100 billion tons. It would fill a cube having each side roughly about four miles long. Not such a modest request after all!

If a living system continues to double in this way, demanding food in the form of raw materials and energy, it will very soon exhaust the resources of its immediate environment. Thus, in a relatively short time the different individuals will have to compete for food. With only a steady supply of food and energy the whole system cannot continue to expand indefinitely; instead, it will reach a steady state. This implies that at that stage each organism will leave, *on the average,* only a single successor in each generation. Since some organisms will double, others must fail to reproduce. This may happen by chance. One organism may light on a local cache of food, while another may be less lucky and starve. However, if one particular organism has acquired a mutation in one of its genes so that for one reason or another it can compete more successfully and, on the average, leave more descendants, then it will increase its representation in the population and thus, necessarily, the other less-favored organisms will producer fewer descendants. If this process continues indefinitely, the less-favored types will eventually die out completely and the one with the more efficient gene will take over completely. The important thing to notice is that by this simple process *a rare chance event has become common.*

The process need not happen only once. It can happen time after time, as chance throws up new favorable mutations. Moreover, improvement can be added to improvement until, given enough time, the process of evolution will produce an organism very finely tuned to its environment. To reach such a perfection of design it needs only mutations produced by chance. There appears to be no mechanism, certainly no common mechanism, to *direct* the change in the gene so that only favorable alterations are produced. Moreover, one can argue that such a directed mechanism in the long run would be too rigid. When times get tough, true novelty is needed—novelty whose important features cannot be preplanned—and for this we must rely on chance. *Chance is the only source of true novelty.*

Such is the power of natural selection that it can operate at all levels. In particular, it can produce improvements in the mechanisms for selection itself—sexual reproduction would be an example of this. If the environment—itself a concept not easy to define precisely—remains stable, natural selection often tends to be conservative and keeps a set of interbreeding organisms within a narrow range, since, in a loose sense, perfection has already been reached and any further improvement may need an exceedingly rare event, all the moderately rare ones having been tried out by that time. However, if the environment alters, or if some individuals become effectively isolated from the rest for some reason or another, then the equilibrium may be upset and under these circumstances natural selection may be more creative. These complexities, which are of major importance for the detailed theory of evolution, need not detain us here, especially as much of our concern will be with the origin of life, when the processes available were probably rather crude. The important thing at this point is to grasp the broad, general features of the process and to realize clearly how such a simple set of assumptions can lead to such remarkable and unexpected results.

As far as we know, there is no other mechanism which can be relied on to produce comparable results so efficiently. One possibility might be the inheritance of acquired characteristics. By striving, a giraffe might make its neck longer and thus obtain more food from the highest, tenderest leaves on the tree. We can conceive that this might cause its offspring to have longer necks and thus be more fit in the struggle for existence. As far as I know, no one has given *general* theoretical reasons why such a mechanism must be less efficient than natural selection, though one would suspect that it may be less flexible than the latter, especially when true novelty is required to surmount an evolutionary crisis. It would in any case demand a process whereby information in the soma (the body of an animal or plant) is conveyed to the germ line— the eggs or the sperm. Such a mechanism has been suggested recently, but the evidence in its favor is complicated and at present rather flimsy. The inheritance of acquired characteristics may conceivably play some small part in evolution, but it is very unlikely that it will turn out to be a major one.

Are there other very general requirements necessary for a living system? For any form of life to be of any serious interest to us it must be at least moderately complicated and probably needs to be very complicated indeed. We know of nothing in the structure of the universe, at any level, which produces such a degree of complexity because of the nature of things. The only mechanism we know that can do this is natural selection, whose requirements we have just outlined.

We have seen that this implies the storing and replication of a large amount of information. The only efficient way to do this is to use the combinational principle. That is, we express the information by using only a small number of types of standard units, but we combine them in very many different ways. (Writing is an excellent example of this principle.) Life as we know it uses linear strings of standard units, but it is possible to conceive schemes which use ordered sheets of units or even structures in three dimensions, though these would be less easy to replicate. Not only must these structures contain information—that is, they must not be completely regular—but their informational content must be easy to copy accurately and, most important, the information must be stable for a much longer time than is needed to copy it, otherwise errors would be too frequent and natural selection could not operate. Thus, the construction from standard units of extended combinations which are fairly stable would seem to be essential if any higher form of life is to evolve. If we try to avoid the use of a small number of standard units the mechanism of replication becomes increasingly difficult, as indeed it is for the printing or typing of Chinese, which contains thousands of different units.

Another general requirement is that the process must not be too slow. We cannot, as yet, calculate the rate of evolution from first principles, but a system which was, say, ten or a hundred times as slow as ours would hardly have had time to produce higher organisms of a complexity similar to ours, even if the system started soon after the Big Bang. Thus, any system based on the solid state, where chemical reactions do indeed proceed but do so extremely slowly, would almost certainly not be fast enough. This leaves us with liquids and gases to consider.

One objection to a purely gaseous state is that only small molecules can form true gases, since even if there are no specific forces of attraction between them there are always appreciable nonspecific forces (called van der Waals forces). These act between all atoms, though only at short distances, and they increase with molecular size. Since, as we have seen, an informational molecule must be rather large (in order to embody instructions using the combinational method), it is unlikely to be gaseous, except at higher temperatures, when it is in danger of being broken into pieces by the thermal motion, or at extremely low pressures, which would cause other difficulties. In particular, the concentrations of the molecules forming the gaseous phase would then necessarily be low and this would slow down the rates of the chemical reactions needed. For all these reasons it is very difficult to devise any plausible system based solely on a purely gaseous phase.

There are more possibilities if we allow specks of solid matter or drops of liquid (or drops surrounded by a special skin) to flow about in a gaseous phase. In such cases it is more difficult to argue that such a form of life is highly unlikely. It might be thought impossible to evolve any large organisms using such a system, but here one must be careful. The very existence of land animals and plants shows that once a system has progressed some way, natural selection can be very ingenious at surmounting obstacles of this kind. Yet however one looks at the problem, the easiest solution is to employ a system based on large combinations, resembling the solid state but on a minute scale, floating about in a liquid. Anything else would seem extremely difficult to get going. As carbon is the atom which, above all others, excels in bonding with other atoms, thus producing an almost infinite variety of organic molecules, and as water is the most abundant molecule in the universe which is likely to be found in any quantity in the liquid state, it is not too surprising that life as we know it is based on carbon compounds in solution in water.

Of course, elsewhere in the universe life may exist based on other materials. At lower temperatures liquid ammonia might serve as the solvent, though it is not as versatile a solvent as water, which is an exceptionally good one. Instead of carbon, silicon has been suggested. This has the advantage that it is fairly abundant, at least on the surface of the earth. Silicon, combined with oxygen to make silicate, does indeed form extended structures. Some of these are sheets, a few are linear, but most are rather intricate three-dimensional structures, crystalline or pseudocrystalline, and do not look as if they could easily form a basis for natural selection, except in a very clumsy manner.

Thus, a form of life based on other materials is not impossible. Some systems deserve further study, but so far no one has succeeded in proposing one which really looks promising. Some systems, such as life in a plasma or life inside a star, appear most unlikely. To achieve a form of life in the interior of the sun it would be necessary to have a large variety of extended combinations of nucleons which were stable for long times. Admittedly, events inside the sun might proceed very rapidly indeed, because the temperature there is so high. (In fact, nuclear reactions there go very slowly, which explains why the sun has been shining so steadily for so long.) Perhaps when a star explodes the reactions might be considered to be a very primitive form of natural selection, but the explosion is so transient that the results would be frozen almost before the process had time to get going.

Fortunately, these rather remote possibilities need not concern us here. Our form of life is clearly based on carbon compounds in a watery medium. What are these organic chemicals like and how do they interact with each other?

FIVE

Nucleic Acids and Molecular Replication

Now that we have described the requirements for a living system in rather abstract terms, we must examine more closely how the various processes are carried out in the organisms we find all around us. As we have seen, the absolutely central requirement is for some rather precise method of replication and, in particular, for copying a long linear macromolecule put together from a standard set of subunits. On earth this role is played by one or the other of the two great families of nucleic acids, the DNA family and the RNA family. The general plan of these molecules is extremely simple, so simple indeed that it strongly suggests that they go right back to the very beginning of life.

DNA and RNA are rather similar—molecular cousins, you might say—so let us describe DNA first and then how RNA differs from it. One chain of DNA consists of a uniform backbone, the sequence of atoms repeating over and over again, with a side-group joined on at every repeat. Chemi-

The base-pairs which are the secret of the DNA structure. The bases are held together by weak hydrogen bonds, shown by the interrupted lines. Thymine always pairs with adenine; cytosine with guanine.

cally the backbone goes . . . phosphate-sugar phosphate-sugar . . . etc., repeating many thousands or even millions of times. The sugar is not the sugar you have on your breakfast table but a smaller one called deoxyribose—that is, ribose with one "oxy" group missing (hence the name DNA, standing for DeoxyriboNucleic Acid—"nucleic" because it is found in the nucleus of higher cells, and "acid" because of the phosphate groups, each of which in normal conditions carries a negative charge). Each sugar has a side-group joined to it. The side-groups differ, but there are only four main types of them. These four side-groups of DNA (for technical reasons called *bases*) are conveniently denoted by their initial letters, A, G, T and C (standing for *A*denine, *G*uanine, *T*hymine and *C*ytosine, respectively). Because of their exact size and shape and the nature of the chemical constituents, A will pair neatly with T, G with C. (A and G are big, T and C are smaller, so each pair consists of one big one with one smaller one.)

Both DNA and RNA rather easily form two-chain structures, in which the two chains lie together, side by side, twisted around one another to form a double helix and linked together by their bases. At each level there is a base-pair, formed between a base on one chain paired (using the pairing rules) with a base on the other. The bonds holding these pairs together are individually rather weak, though collectively they make a double helix reasonably stable. But if the structure is heated the increased thermal agitation will jostle the chains apart, so that they separate and float away from each other in the surrounding water.

The genetic message is conveyed by the exact base-sequence along one chain. Given this sequence, then the sequence of its complementary companion can be read off, using the base-pairing rules (A with T, G with C). The genetic information is recorded twice, once on each chain. This can be useful if one chain is damaged, since it can be repaired using the information—the base-sequence—of the other chain.

There is one unexpected peculiarity. In the usual double helix the two backbones of the two chains are not approximately parallel but antiparallel. If the sequence of the atoms in one backbone runs up, that in the other runs down. This does cause certain complications, but not as much as one might expect. At bottom it springs from the type of symmetry possessed by the double helix. This is produced by the pseudosymmetry of the base-pairing. It happens to be the convenient way for these particular chemicals to fit neatly together.

It is easy to see that a molecule of this type, consisting of a pair of chains whose irregular elements (the bases) fit together, is ideal for molecular replication, especially since the two chains can be rather easily separated from each other by mild methods. This is because the bonds *within* each chain, holding each chain together, are strong chemical bonds, fairly immune to normal thermal battering, whereas the two chains cling to each other by rather weak bonds so that they can be prized apart without too much difficulty and without breaking the individual backbones. The two chains of DNA are like two lovers, held tightly together in an intimate embrace, but separable because however closely they fit together each has a unity which is stronger than the bonds which unite them.

Because they fit together so precisely, each chain can be regarded as a mold for the other one. Conceptually the basic replication mechanism is very straightforward. The two chains are separated. Each chain then acts as a template for the assembly of a new companion chain, using as raw material a supply of four standard components. When this operation has been completed we shall have two pairs of chains instead of one, and since to do a neat job the assembly must obey the base-pairing rules (A with T, G with C), the base-sequences will have been copied exactly. We shall end up with two double helices where we only had one before. Each daughter double helix will consist of one old chain and one newly synthesized chain fitting closely together, and more important, the base-sequence of these two daughters will be identical to that of the original parental DNA.

The basic idea could hardly be simpler. The only rather unexpected feature is that the two chains are not identical but complementary. One could conceive an even simpler mechanism in which like paired with like, so that the two paired chains were identical, but the nature of chemical interactions makes it somewhat easier for complementary molecules, rather than identical ones, to fit together.

what

How does such a process compare with the grosser copying mechanisms commonly used today? A line of type, made up for printing, consists (or used to consist) of a set of standard symbols arranged in a line or a series of lines. Each letter from the font has a standard part, the same for all letters, which fits into the grooves which hold the type in place, and a part which is characteristic of each letter. After that the resemblance ceases. There is nothing in DNA replication which corresponds to the ink. The letters printed on the page are the mirror images of the typeface, not the complement (which would stick out when the typeface went in), and, most important, the resulting line of print cannot then be put back into the same machine to reproduce the typeface. Printing presses produce many thousands of copies of newspapers, but newspapers are not copied back into type.

DNA replication is not like that. For natural selection to work it is essential that the copy can itself be copied. DNA replication is more like production of a piece of sculpture from a mold, since if it is sufficiently simple the sculpture can itself be used to produce a further mold. The main difference is that a strand of DNA is built from just four standard pieces. This is obviously not true of most pieces of sculpture.

If we examine the process of DNA replication, we see that

there are a number of basic requirements. If we start with a double helix, the two chains must be separated in some way. There must be available a supply of the four components. Each of these consists of the relevant piece of the backbone —one sugar molecule joined to one phosphate—plus one of the four bases attached to the sugar. Such a tripartite molecule is called a nucleotide. In practice these precursors have not just one phosphate but three in a row, the other two being split off in the process of polymerization, thus providing the energy to drive the synthesis in the desired direction. Though one can conceive of the process proceeding without extra components, in an evolved system we would expect to find at least one enzyme (a protein with catalytic activity, that is) which would accelerate the synthesis and make it more accurate.

Such are the requirements in outline. When a real replicative system is examined it is found to be considerably more elaborate. To begin with, the two chains are not first completely separated before synthesis starts. Synthesis of the new chains proceeds during the process of separation, so that some parts of the double helix have been replicated before other more distant parts have been separated. There are special proteins whose job it is to unwind the double helix, together with others which can put nicks in the backbone, to allow one chain to rotate around the other, and then join up the broken chain again. Since the two chains of the double helix run in opposite directions, and since, chemically speaking, the synthesis goes in only one direction, we find that synthesis is directed forward on one chain and backward on the other, so the mechanism has to allow for this complication. Moreover, a new fragment of a DNA chain is usually started as a small length of RNA, to which a longer piece of DNA is then joined. There are additional proteins which then cut out this RNA primer and replace it with an equivalent bit of DNA chain and then join everything together without a break. To synthesize one particular small virus made of DNA we know that almost twenty distinct proteins are required, some to do one job, some to do another. This is very characteristic of biological processes. The underlying mechanism may be simple, but if the process is biologically important, then, in the long course of evolution, natural selection will have improved it and embroidered it, so that it can work both faster and more accurately. It is because of this baroque elaboration that biological mechanisms are often so difficult to unravel.

Fortunately, as we noted earlier, these complications need not detain us. When life started, the chemistry must have been relatively simple. The important point to grasp is that the neat geometry of the base-pair, which underlies the pairing rules, gives the opportunity of specific replication, even in quite unsophisticated systems. We see that the crucial thing about DNA is not that it is a double helix. In fact, a simple virus may have a single strand of DNA as its genetic material, as it may be so short (a mere five thousand bases long) that it does not require the second chain as an insurance against damage. The essential feature is that the replicative mechanism should use the simplicity of the specific base-pairs to build a new chain with a base-sequence complementary to the old one. It is this simplicity which tempts us to believe that it was used in the very earliest living systems. Whether the two chains—the old and the new—stay together after the replication is a matter of less importance.

At this point we must say a few words about DNA's close relative, RNA. (The various types of RNA are described more fully in the Appendix.) As we have explained, the genetic information in each cell of a higher organism is en-

coded as the detailed base-sequence of a number of very long
DNA molecules. At any one time many shorter parts of this
sequence are being copied onto single-stranded RNA molecules to be used as working copies by the cell. Some of these
are used for structural purposes, but most of them are used
as messenger RNA, the instructions for protein synthesis.
This occurs on very complex molecular structures called ribosomes and needs a lot of auxiliary molecular apparatus, in
particular, a set of tRNA molecules.

The system is undoubtedly very complex, but this is
mainly because a complicated job has to be done. The process
of making a single-stranded RNA copy of a stretch of DNA
—called transcription—is relatively straightforward and
only requires a rather large protein to direct it. The process
of synthesizing a protein using a piece of messenger RNA as
instructions—called translation—is necessarily more difficult, since the instructions are embodied in the four-letter
RNA language but have to be *translated* by the chemical
machinery into the twenty-letter protein language. Indeed,
it is quite remarkable that such a mechanism exists at all and
even more remarkable that every living cell, whether animal,
plant or microbial, contains a version of it. Its elucidation
has been one of the triumphs of molecular biology.

The cell is thus a minute factory, bustling with rapid,
organized chemical activity. Under suitable molecular con-
trols, enzymes busily synthesize lengths of messenger RNA.
A ribosome will jump onto each messenger RNA molecule,
moving along it, reading off its base-sequence and stringing
together amino acids (carried to it by tRNA molecules) to
make a polypeptide chain which, when finished, will fold on
itself and become a protein. Nature invented the assembly
line some billions of years before Henry Ford. Moreover, this
assembly line produces many different highly specific pro-

teins, the machine tools of the cell, which themselves shape and reshape the organic chemical molecules in order to provide raw material for the assembly lines and also all the molecules needed to build the structure of the factory, provide it with energy, dispose of the garbage and a host of other functions. Because it is so complicated the reader should not attempt to struggle with all the details. The important point to realize is that in spite of the genetic code being almost universal, the mechanism necessary to embody it is far too complex to have arisen in one blow. It must have evolved from something much simpler. Indeed, the major problem in understanding the origin of life is trying to guess what the simpler system might have been.

At this stage it is perhaps worthwhile to compare and contrast these three great families of macromolecules: protein, RNA and DNA. Protein molecules, being constructed from twenty different side-chains, several of which are chemically rather active, are much more versatile as a class than the nucleic acid molecules. It is for this reason that all known enzymes are made of proteins, though in certain cases a small organic molecule may be needed to work with it as a coenzyme. It is the ability of each enzyme to make or break particular chemical bonds which allows modern cells to function at all. Since many different chemical reactions need to be catalyzed in this way, there are many different kinds of enzymes.

By contrast, no nucleic acid molecule has been found with any catalytic activity. Both RNA and DNA have only four types of side-groups instead of twenty, and although ideal for replication because the bases fit together so well, these side-groups would not be very good for chemical catalysis. But RNA and DNA can do what proteins cannot do—form complementary structures of the type found in the double helix.

We know of no way in which a protein molecule could do this, certainly not a modern protein with its twenty different kinds of side-chains.

Most chemists working on the origin of life suspect that in the beginning RNA came first and that DNA was a later invention. RNA is chemically more reactive than DNA and it was probably easier to synthesize under primitive earth conditions. The very earliest genes may have been made of RNA. Only later, when the genetic information grew in length, was the more stable DNA needed to provide the file copy.

Life, as we know it on earth, appears as a synthesis of two macromolecular systems. The proteins, because of their versatility and chemical reactivity, do all the work but are unable to replicate themselves in any simple way. The nucleic acids seem tailor-made for replication but can achieve rather little else compared with the more elaborate and better equipped proteins. RNA and DNA are the dumb blondes of the biomolecular world, fit mainly for reproduction (with a little help from proteins) but of little use for much of the really demanding work. The problem of the origin of life would be a great deal easier to approach if there were only one family of macromolecules, capable of doing both jobs, replication and catalysis, but life as we know it employs two families. This may well be due to the fact that no macromolecule exists which could conveniently carry out both functions, because of the limitations of organic chemistry; because, that is, of the nature of things.

To make any further progress we must try to learn something of the chemical and physical conditions on the primitive earth, or on any similar planet. To this we now turn.

SIX

The Primitive Earth

WHAT SUBSTANCES do we need to form the material basis of
life? The life we see all around us is based on carbon atoms,
combined with hydrogen, oxygen and nitrogen, together
with some phosphorus and sulphur. Using these few types of
atom it is possible to construct an enormous number of different small molecules—that is, molecules with less than,
say, fifty atoms—and an almost unlimited number of different macromolecules, each containing thousands of atoms.
Other atoms are also important, such as the charged atoms
(ions) of sodium, potassium, magnesium, chloride, calcium,
iron and a number of others, but in most cases these do not
form part of the organic molecules but exist mainly on their
own. For life to have got started a supply of most of these
atoms was needed. Where did they come from? And were
they solitary or in simple combinations?

It happens that the atoms found in organic chemistry are
all very reactive. Even in the atmosphere they exist com-

bined. Straightforward chemical arguments suggest that hydrogen will combine with itself to form the molecule H_2, oxygen to make O_2 and nitrogen to make N_2. We may also expect simple combinations such as H_2O (water), NH_3 (ammonia), CO_2 (carbon dioxide), CH_4 (methane) and a number of others. Our atmosphere today consists mainly of the very inert gas nitrogen (N_2), together with about twenty percent oxygen (O_2) with a little water vapor (H_2O) and even less carbon dioxide (CO_2).

It used to be thought that the primitive atmosphere on the earth was quite different. Since hydrogen is by far the most abundant element in the universe, it was natural to believe that hydrogen dominated the primitive atmosphere. At the present time almost all the oxygen in the air is produced by photosynthesis. At the earliest times there was no life on earth and so no oxygen could have been produced in this way. Such an atmosphere, rich in hydrogen and poor in oxygen, is known as *reducing,* as opposed to our present atmosphere, which is called *oxidizing.* The experiments on pre-biotic synthesis, to be described shortly, appeared to support this conclusion.

Recently these ideas have been questioned. Hydrogen is so light that the earth's gravity is not strong enough to hold it and it escapes into space rather easily. The exact rate depends on a number of factors, especially the temperature in the upper atmosphere, since the higher the temperature the faster the atoms or molecules move and the more easily they escape into space. It now seems possible that much of the original hydrogen escaped so quickly that the atmosphere was never dominated by it.

But what about the oxygen? It could not have been produced by photosynthesis, but is there some other plausible mechanism? There was almost certainly plenty of water on the primitive earth and in particular in its atmosphere. Under favorable circumstances ultraviolet light can split water into its component elements. If the hydrogen so produced then escaped into space, the oxygen left behind would have accumulated, and if the process was on a sufficiently big scale the atmosphere might have become rich in oxygen. Today, because of the detailed structure of the present atmosphere, this process does not produce oxygen at an appreciable rate, but it is at least possible that in the remote past conditions were so different that oxygen was produced more freely.

Of course, oxygen and hydrogen were not the only elements present in the air. There was probably a lot of nitrogen, some carbon and perhaps a little sulphur, though the latter two would not have been uncombined. The gases N_2 and CO_2 were probably present, together with smaller amounts of CH_4, CO and perhaps NH_3 and H_2S (hydrogen sulphide). What is quite unclear is their exact proportions, in particular the amounts of H_2 and O_2.

Since the atmosphere interacts with the chemicals on the surface of the earth, the chemical composition of the earliest sedimentary rocks should give us some clues to the composition of the early atmosphere. Some of those rocks suggest that they were formed under reducing conditions. This was taken to support the hypothesis that the atmosphere then was reducing. This also has recently been called into question. Even today some sediments are reducing—stinking muds, for example—in spite of all the oxygen in the air around us. Such conditions are usually produced by the anerobic decay of organic materials in the mud. It is now claimed that if *all* the available rocks of a given age are considered, then, when averaged, the evidence suggests that the atmosphere in the past was rather like what it is today. Unfortunately, this only takes us back to 3.2 billion years ago. Before that the evidence is too sparse, because too few suitable rocks are available to us. The deduction that the atmosphere 3.2 billion years ago was not reducing is not totally surprising, because we believe that there were photosynthetic organisms at least as early as 3.6 billion years ago. Unfortunately, we cannot at the moment discover how many of them there were, so it is difficult to estimate whether their oxygen production was large or small.

In summary, we would like to know the approximate composition of the atmosphere on the earth at a time before life was present, and in particular just how reducing or oxidizing it was. At the moment it seems very difficult to come to any firm conclusion on this matter.

The temperature of the primitive earth is equally uncertain, since this depends largely on how rapidly it formed. If it fell together in a short space of time, the heat generated by the collisions would not have had time to escape, so that at the early stage the earth would have been very hot. If the process was slower, the primitive earth may have had a more moderate temperature, though there would have been transient local hot spots due to impacts during the final stages of the aggregation. Whatever the details of the process, it seems likely that at some point the earth settled down with sufficient liquid water to form the primitive oceans, seas, rivers, lakes and pools.

Whatever the nature of the atmosphere, it was undoubtedly the recipient of a large flux of energy from the sun. It is not known for certain just how hot the sun was at that time, though it is possible that its radiation was not greatly different from what we receive today. One possible difference affecting the radiation which reached the surface of the earth may have been the absence of the present ozone (O_3) layer, since if there was little oxygen in the atmosphere (except that combined as water, CO and CO_2) the ozone layer would have been absent. This layer today screens off much of the ultra-violet light coming from the sun. There were probably, as today, frequent electrical storms (similar to our thunder-storms) and probably a fair amount of volcanic activity, both on the earth and under the oceans. In addition, there were ion-molecule reactions in the ionosphere and upper atmosphere, so that there were several sources of energy of the sort necessary to promote chemical change. All this suggests that the primitive oceans did not consist merely of water and a few simple salts but had accumulated a fair variety of small organic molecules, formed from the molecules in the atmosphere and dissolved in the oceans by means of electrical discharges, ultraviolet light or other energy sources.

The idea that the early atmosphere was not like the present one but contained much less oxygen appeared to receive dramatic support in 1953 from Stanley Miller, a student of Harold Urey, who passed an electrical discharge through a mixture of CH_4, NH_3, H_2 and H_2O contained in a closed system. The system included a flask of water which was boiled to promote circulation of the gases and which served to trap any volatile water-soluble products which were formed and protect them from dismemberment by the electric spark. After a week or so the discharge was stopped. The water was found to contain a variety of small organic compounds, including a fair amount of two simple amino acids, glycine and alanine, found in all proteins. Many similar experiments have since been done, using different mixtures of gases and a variety of sources of energy and experimental conditions, including passing the gases over heated mineral surfaces. The results are too complex to summarize here except for one striking fact. If the mixture of gases contains

appreciable amounts of oxygen, then small molecules related
to molecules present in living systems are not found. If gaseous oxygen is absent, such small molecules are produced,
provided the mixture of gases contains nitrogen and carbon
in some form or other. Some gas mixtures produce a bigger
variety of amino acids than others, especially if H_2 is not
present. On the primitive earth, H_2 would have been lost
into space, whereas in Miller's original experiment, which
was in a closed vessel, any H_2 formed had no similar way of
leaving the apparatus and thus accumulated as the experiment went on.

Thus, if the atmosphere was reducing it is likely that the
water on the primitive earth contained a rather dilute mixture of small organic molecules, many of which might serve
as raw materials for the earliest living systems. Exactly which
molecules were formed, and in what quantity and where—
whether in the upper atmosphere, in the oceans, near sub-
marine volcanos, or in tidal pools, small lakes, hot springs,
near volcanic crevices or in all these places—is open to debate. Many of these molecules are not stable in water over
very long periods of time, so that eventually the amounts
found would be due to a balance between their continued
production over thousands or millions of years and their de-
struction in the water due to thermal motion. Most amino
acids have both a negative charge and a positive one, so that
although they are small and, in sum, electrically neutral,
they prefer to stay in the water rather than escape into the
air. For this reason they would not have been lost by evapo-
ration. Nor would this primitive soup, as it is often called,
have "gone bad" in the ordinary sense, because there were
then no microorganisms to live in it and use its molecules for
food.

I once asked my colleague Leslie Orgel, who works on the

origin of life, how concentrated this soup might have been. He told me that he had done a very rough calculation and that it probably contained about as much organic matter (though mainly in small organic molecules) as chicken soup. I was flabbergasted. I distinctly remembered that on one of those rare occasions when I had to cook my own dinner I had opened a tin of chicken soup and that, apart from small chunks of meat, it was a thick, rich, creamy mixture. A whole ocean of that seemed to me to be highly unlikely. However, it turns out that this material is more correctly described as chicken *broth*. What Orgel had in mind was a clear, rather thin, chicken bouillon. He had, in fact, gone so far as to measure the amount of organic material in a particular sample of it. Perhaps not everybody would agree with his estimate, but it does give a very rough idea of the total amount of organic raw material which was probably available on the earth before life began.

If it turns out that the early atmosphere was not reducing but contained a fair amount of oxygen, then the picture is more complicated. At first sight it might seem that since no suitable raw material was available, life could hardly have got started here. If this were really true, it would support the idea of Directed Panspermia, because planets elsewhere in the universe may have had a more reducing atmosphere (as we shall discuss in Chapter 8) and thus have on them a more favorable prebiotic soup. However, even under an oxidizing atmosphere there may have been some places on the earth where conditions were reducing—under rocks and at the bottom of lakes and oceans, for example. Perhaps there were hot springs on the sea floor which provided around them suitable conditions for prebiotic synthesis.

Another possibility is that appreciable amounts of the small molecules found in space reached the earth's surface by one mechanism or another, perhaps on comets which collided with it, producing local concentrations of suitable chemicals. Even if they only amounted to a small fraction of the earth's surface, there may have been enough of these special places to get things going, assuming that life can start very easily, given the right environment.

In spite of all these uncertainties, it seems possible that in some early stage in the earth's history there was a fair amount of water on its surface, and that in such places it consisted of a weak solution of small organic molecules, many of them not unrelated to the raw materials needed to construct proteins and nucleic acids, together with various salts washed out of the surrounding rocks. The conditions might well have been suitable for the emergence of some very primitive form of life. We are thus faced with the difficulty of deciding at what stage in this continuing process of chemical evolution we should accept such a very simple system as living.

The selection of any particular stage must be to some extent arbitrary, but there is one criterion we can usefully apply to make the demarcation between the living and the nonliving. Is natural selection operating, even if only in a rather simple way? If it is, then a rare event can be made common. If not, any rare event must be solely due to chance and the intrinsic nature of things. This criterion is important because, as we shall see, the origin of life may indeed have been a rare event and we would very much like to know exactly how rare it was.

How likely was it, given a soup of one sort or another, that a system arose spontaneously which could evolve by natural selection? Here we face formidable problems. Whatever happened during those early times, we can be sure that the primitive system had eventually to evolve fairly smoothly into the present one, based on nucleic acid for replication and protein synthesis for action. We cannot be sure that the earliest evolving system was not embodied in something quite different, which set the stage for the present one. Even if this was not the case, and the first replicating system contained some elements of the one we have today, we have no evidence whether nucleic acid came first, or protein came first, or whether both evolved together. My own prejudice is that nucleic acid (probably RNA) came first, closely followed by a simple form of protein synthesis. This seems to me the easiest route to follow, but even this appears fraught with difficulties. Phosphate was probably common and the sugar ribose (which contains no nitrogen) could have easily been made under certain special conditions, because formaldehyde ($HCHO$) is known to be one of the most common prebiotic chemicals. However, a rather different set of conditions would have been required for the synthesis of the bases, such as adenine, which do contain nitrogen. Then there is the problem of linking the sugar to both the phosphate and the base in the correct way (and several incorrect ways are possible) and then activating this compound (called a nucleotide), possibly by joining on a further phosphate or two to provide the energy needed to link two nucleotides together. This operation, if repeated, would lead to the chain molecule we call RNA. It is not easy to see how this could happen in a mixture of other, rather similar compounds without the frequent incorporation of incorrect molecules in the chain unless there were some rather specific catalyst present. This conceivably could be a mineral or even some peptide produced by the random aggregation of amino acids, but if so this has not been demonstrated in a convincing way. Even if such a process did occur, if only in one particular pool at one particular time, it would only yield RNA with a rather random base-sequence.

For natural selection to operate we need a reasonably accurate copying mechanism. Here we can see a gleam of hope.
If RNA polymerization were for some reason rather common,
it would very likely have led in time to some molecule similar
to the tRNA molecules used universally in present-day protein synthesis. The loops of such a molecule might help to
condense nucleotides into short chains only three residues
long, and these might be better precursors for a replication
process than single nucleotides.

If replication were all that was needed, RNA would seem
a very promising candidate, but although replication alone
may get a system going, something more is required as competition increases. Before too long a gene must do something
if it is to have much impact on its surroundings. Now, RNA
is not ideal for this. It can indeed, in favorable cases, form
three-dimensional structures, but these seldom seem to have
any catalytic activity. Perhaps this was provided by small
organic molecules of one sort or another, abundant in the
surrounding soup. Some of these may have combined neatly
with certain folded RNA molecules to produce a primitive
"enzyme" with a small amount of rather crude catalytic activity, though so far nobody has attempted to discover such
entities.

A more attractive alternative is that a primitive system of
protein synthesis might have started with a messenger RNA
molecule and tRNA alone, that is, without ribosomes or
protein. This again is a distinct possibility but one not as yet
supported by experiment. Such a system, if it worked, would
get over most of our conceptual difficulties, though some
problems remain—how to attract the "correct" amino acid
to each type of tRNA molecule, for example.

Once RNA synthesis and replication had got going one
might expect that simple catalysts would be produced which

would make all these early chemical reactions more rapid and more efficient. From then on natural selection could operate to refine and develop the system. Attractive though this is, no method of doing it has yet been worked out in all its details and tested experimentally.

There is thus some reason for looking at other alternatives. A second obvious candidate for the primitive replicator is some kind of early protein. This is attractive because the soup almost certainly contained some amino acids, and possibly a fair number of different types, though (apart from glycine, which has no hand) there would have been roughly equal mixtures of the two possible hands. The difficulty here is that the amino acids do not appear to pair up neatly the way the bases can pair in the nucleic acids. No double helix of protein has been discovered, though the protein collagen (of tendons, skin, leather, etc.) consists of three polypeptide chains wound around each other to form a triple helix. Every third residue must be glycine, but there seems to be no obvious interaction which might select the amino acids for the other two places. Moreover, collagen has a rather regular structure and appears catalytically inert. If someone could produce a simple form of protein, made of perhaps four amino acids, which could form the basis of a simple copying process (as RNA or DNA can), this would be a major discovery. Till then the claim that protein was the primitive replicator must be viewed with reserve.

This does not mean that accidental polymerization might not have produced proteinoid molecules which might perhaps have assisted in the buildup before true replication finally occurred, but it is this later process which was needed if natural selection was to have operated freely.

There is always the possibility that the early replicating system was of some quite different form which, because it

was too clumsy or not versatile enough, was eventually discarded in place of the present one. Such an idea is difficult to
refute. We should at least be able to imagine how the
changeover was made from the early system, whatever it was,
to the present one based on nucleic acid and protein. It has
been suggested that layered clay structures might be suitable,
but it is not easy to see in detail how they would have
worked, and no dramatic experimental evidence for such behavior has so far been produced.

All in all, it seems rather plausible that the first replicator
was RNA. This hypothesis would gain considerably if we
could put together a simple copying system in the test tube,
using no protein. To make it easy we might start with one
preformed strand of RNA, having some arbitrary base-sequence, and try to make its complementary companion by
supplying the necessary raw materials. We should need four
types of these and some form of chemical energy to drive the
reaction. Such experiments have been done. So far they have
only met with rather modest success. The best performance
to date, by Leslie Orgel and his colleagues, was poly C
(polycytidylic acid) as a template—that is, an RNA every
base of which is cytosine—supplied with a chemically acti-
vated form of G, the normal complement of C. In the presence of zinc ions (Zn^{++})—an ion found in all present
enzymes which polymerize nucleic acid—the Gs are slowly
joined together in the correct linkage (called $3'-5'$) to make
poly G of appreciable length. Molecules as long as forty Gs
in a row can be detected in the incubation mixture, and
longer ones are likely to be present in amounts below the
present levels of detection. Moreover, the system is reasonably accurate in that only rather small amounts of A and U
are incorporated (as "errors") when their precursors are also
added to the mixture. This is a promising beginning, but for

it to be useful one should be able to see the exact (complementary) replication of a particular sequence of Cs and Gs. So far this has not been done. It is, incidentally, not essential to have all four bases in the original system, since an RNA with only two types of them can embody information in its sequence; however, for good replication the two must be complementary.

Even if these difficulties are overcome, the system, though simple, is already somewhat sophisticated. It is, for example, unnaturally pure. It is difficult to imagine how a little pond with just these components, and no others, could have formed on the primitive earth. Nor is it easy to see exactly how the precursors would have arisen. These might be expected to be nucleoside triphosphates—in simpler terms, molecules consisting of a base, a sugar (ribose) and three phosphates in a row, though those were not the exact compounds used in the experiment described above. It is possible to see how each of these separate components might possibly have arisen on the primitive earth in one place or another; it is less easy to see how the combination was formed correctly and how it was at least partially separated from other, rather similar molecules which, if present, might possibly have fouled up the system. Certainly nobody has been able to cook up a primitive soup with water, salts, a few gases and ultraviolet light (or some other energy source) and let it stew away till a neat RNA replicating system arose from it. This failure is not too surprising, since it may have taken nature many millions of years, in many places on the earth's surface, before one happy combination of circumstances produced a system which could both initiate replication and also keep going for some time.

We are thus in a most tantalizing situation. On the one hand we believe that there may have been a fairly adequate

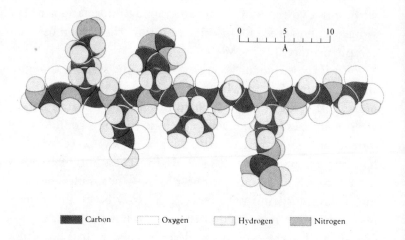

	Carbon		Oxygen		Hydrogen		Nitrogen

The model represents a short, extended polypeptide only nine amino acids long. The backbone of the chain is regular, with side-groups attached at regular intervals.

supply of organic molecules, amino acids in particular, on the earth's surface, even though their concentration in most places may have been somewhat low. In addition, the double helix of RNA or DNA certainly suggests that it could form a good basis for a primitive replication system. On the other hand, it is difficult to see how an accurate system could have arisen easily from such a complex mixture, and even more difficult to figure out the exact components needed and the exact steps followed. Moreover, even if one could see how

RNA replication could have started, we have yet to work out how it became coupled to even a primitive form of protein synthesis, although we can begin to make some educated guesses as to how this might have happened.

What is so frustrating for our present purpose is that it seems almost impossible to give *any* numerical value to the probability of what seems a rather unlikely sequence of events. The difficulty can be seen more clearly by the following very crude argument. Let us suppose that the event took place in some pond or pool, perhaps near the margins of the sea. We could easily imagine that there was such a pool every mile or so of coastline, to say nothing of those sprinkled over the surface of the earth. Perhaps 100,000 such places existed —the number could easily be much higher. Again let us postulate that at the slow rate at which such systems work it might take a time like a hundred years for one to get going. Let the very small probability of such an event happening in a hundred years be called p. Perhaps p was one in a billion. But since we have perhaps 500 million years and 100,000 pools, we see that in that case life was almost certain to have got started. However, if p was only one chance in a billion billion, the chance of starting was not far from even. If as little as one in 10^{15} (a thousand billion billion), the chance of life starting here was very small. The exact figures do not matter all that much. They are merely used to show the type of dilemma involved. This springs from the fact that we have no idea what value we should take for p, except that it should be "small." For this reason it is impossible for us to decide whether the origin of life here was a very rare event or one almost certain to have occurred. Even though arguments are sometimes put forward for the latter view, they seem very hollow to me. Without some direct experimental support they are likely to remain so. And to get experimental support for what could well have been a sequence of fairly rare reactions is not going to be easy. Only if life was *very* easy to start, because there is in fact some rather direct pathway through the maze of possibilities, are we likely to be able to reproduce it in laboratories, at least in the immediate future.

An honest man, armed with all the knowledge available to us now, could only state that in some sense, the origin of life appears at the moment to be almost a miracle, so many are the conditions which would have had to have been satisfied to get it going. But this should not be taken to imply that there are good reasons to believe that it could *not* have started on the earth by a perfectly reasonable sequence of fairly ordinary chemical reactions. The plain fact is that the time available was too long, the many microenvironments on the earth's surface too diverse, the various chemical possibilities too numerous and our own knowledge and imagination too feeble to allow us to be able to unravel exactly how it might or might not have happened such a long time ago, especially as we have no experimental evidence from that era to check our ideas against. Perhaps in the future we may know enough to make a considered guess, but at the present time we can only say that we cannot decide whether the origin of life on earth was an extremely unlikely event or almost a certainty—or any possibility in between these two extremes.

If it was highly likely, there is no problem. But if it turns out that it was rather unlikely, then we are compelled to consider whether it might have arisen in other places in the universe where possibly, for one reason or another, conditions were more favorable.

SEVEN

A Statistical Fallacy

summary — just because life exist on earth - does that imply that life must exist on some other planet - no.

IN SPITE OF OUR UNCERTAINTY about how life began, we have no doubt that it exists now, and on an abundant scale. We can see it all around us. Surely, one might argue, since it has happened once we can be confident that it could happen again. Of course, it is most unlikely that it would start again now. Apart from the present conditions being so different from the prebiotic ones, it would seem highly likely that any new system which tried to get going at this time would be gobbled up by members of the existing one. This point of view is a relatively recent one. Even as late as the nineteenth century it was believed that life could arise *de novo,* here and now, in swamps, infusions, rotting meat and other suitable places. There are frequent reports of maggots, flies and even mice originating in this way. The early experiments of Redi, Joblot and Spallanzani made this rather doubtful and the careful and elegant work of Pasteur showed that all such claims were almost certainly false. By the ingenious design of his apparatus Pasteur removed, one by one, all the objections his critics could think of. He showed beyond doubt that in an initially sterile system, no sign of life would appear in even the richest and most tempting brew, even if there were free access to the air, provided care was taken to prevent any microorganisms in the air from reaching the culture vessel.

The questions we are concerned with are rather different ones. If the earth started all over again (with only small variations so that events would not repeat themselves exactly), would we expect to see life beginning for a second time? More to the point, if a planet rather similar to the earth exists elsewhere, what are the chances that life could get going there? Even in these cases there is a strong psychological urge to believe that such events must be highly likely because of the example of life on earth. Unfortunately, this argument is false. I do not know whether such a line of reasoning has a name, but it might be called the *Statistical Fallacy.* The easiest way to understand why it is wrong is to consider any rather well-defined event which is one of a very large number of rather similar possibilities. A pack of cards provides an excellent model for such a situation.

Let us have a conventional pack of fifty-two cards, well shuffled and dealt at random into four hands of thirteen cards each, with no funny business. What is the chance that a particular set of four hands will be dealt? We might choose one which is easy to specify, such as the first hand containing all hearts, the next with all diamonds, the third with all spades and all the clubs in the last hand, but it makes no difference to the calculation, provided we specify exactly what cards we require in each of the four hands. It is a simple matter to calculate how often such a hand will turn up if the cards are dealt, over and over again, from a randomized pack.

The chance of getting it turns out to be as little as one in 5×10^{28}. Yet every time we deal a pack, we get some distribution or other in the four hands, and since our calculation would apply just as well to *that* distribution, such a hand should be exceedingly rare. And yet there it lies on the table before us. Clearly something must be wrong.

What is wrong is that for the calculation to apply, we must say *in advance* exactly what set of hands we are considering. We are not allowed to deal the cards and *then* pretend that the result was just what we were looking for. We can, of course, deal one set and then decide that that is the combination we are going to select. The low probability we calculated would then give the chance of getting this same hand on the next deal, always assuming that the pack had been randomized properly. This argument could apply to whatever the actual hand was, so that we see another way of looking at this figure of one in 5×10^{28} is to say that it is the chance of dealing *any* set of four hands *twice in succession.*

We can put this in a different manner by saying that the composition of *one* deal tells us, in itself, practically nothing about the chance of getting exactly the same four hands again. It does tell us that we have the correct set of fifty-two cards in the pack, but it does not tell us, by itself, what the chance is that we shall get the same set dealt on succeeding occasions. We can calculate this if we know all the parameters in the situation and the number of hands we are going to deal before we give up. This we can know for the pack of cards, but the prebiotic situation is more complicated. There is also the additional factor that we are usually not trying to calculate the chance for an *identical* event to occur a second time. Any reasonable form of life fairly similar to the present one would be acceptable and would count as success. An analogy from the cards may make this clearer. We asked for the four hands to be all hearts, diamonds, spades and clubs *in that order.* But suppose we were to accept as a success any deal in which each hand had cards of only one denomination. This is twenty-four times as likely to occur as for the case we considered above, since there are more possible deals which will satisfy our conditions. When we consider the origin of life, this factor—the number of similar but not identical forms of life—is also quite unknown and only adds to our uncertainty.

In some ways this last general point is at the root of our problem. Even though the probability of life starting at one particular time in one particular place looks exceedingly minute, there were so many possible places on the earth and so much time available that we cannot be sure that these factors do not overwhelm the small probability of any one of them occurring, thus turning a rare event into one which is almost certain. But a moment's thought shows that we have no factual basis for this conclusion. As discussed in the last chapter, the overall probability could be anything, depending on just how the various numbers turned out.

There is a special reason why the Statistical Fallacy applies with particular force in our own case. This is because if life had *not* started here (in one way or another), we would not be here to think about the problem. The mere fact that we *are* here necessarily implies that life *did* get started. For this reason, if for no other, we cannot use this fact directly in our calculations.

We seem to be in contact with an inherent failure of the human mind when confronted with probability arguments. Human beings, and probably other animals as well, are far too prone to generalize from one instance. The technical word for this, interestingly enough, is *superstition,* though many forms of superstition also have an emotional component. We also have trouble grasping very large numbers, so that we are happy if a very small number multiplied by a very large number comes out to something we are more at home with, such as a probability near one. Certainty is often very close to our hearts, however much it eludes us in practice. The only way to overcome these psychological handicaps —and in scientific matters they are indeed handicaps, however useful they may have been in evolution—is to set out the argument coolly and clearly. A "gut reaction" may be useful in business or politics or in our personal lives because it represents an unconscious generalization of previous experience, either our own or that of our ancestors expressed in our genes, but in considering the origin of life we have effectively no experience to guide us in this way, so that any gut reaction is likely to be superficial and misleading. It is even less useful in approaching the chance of life having evolved independently elsewhere. We have rather little knowledge of the planets of our own solar system and none at all, except by very indirect inference, of the planets circling around other stars. Perhaps there are many places in the universe suitable for the origin of life, some of which may have conditions even more favorable than those we find here. It is to these problems which we must now turn.

EIGHT

Other Suitable Planets

OUR MAIN CONCERN is with life as we see it here, based on carbon compounds dissolved in water. We are confronted with a universe of vast extent, mainly empty, but with occasional special places suitable for a form of life not unlike ours. How many such places are there likely to be?

Perhaps the most restrictive requirement is that there should be liquid water. Water itself is likely to be a fairly common compound, but we must have it in an environment which is not so cold that the water will exist only as solid ice and not so hot that it will all be vaporized. The problem can be seen most clearly by expressing temperature in degrees Kelvin—the so-called absolute scale. This is based on the normal centigrade or Celsius scale, in which, under standard pressure conditions, pure water freezes at 0°C and boils at 100°C. On the absolute scale this difference is still 100°, but 0° is taken to be the absolute zero of temperature—loosely speaking, the temperature at which all random motion

ceases. On such a scale ice melts at about 273°K and water
boils one hundred degrees higher, at about 373°K. We must
set these two numbers against the desolate frigidity of space,
whose temperature is, very roughly, 4°K, only a little above
absolute zero, and the temperature at the sun's surface,
which, in round figures, is 5,000°K. Since we need a temperature in the region of 300°K, we see immediately that we
shall only find this fairly close to a star, but not too close.
Most of the universe will not only be too empty but also too
cold. The above very simple argument assumes that the pressure of the gas above the water will be rather like the atmospheric pressure we have at the surface of the earth. If the
pressure were higher we could tolerate a slightly higher temperature and still have liquid water, though the pressure only
alters the allowed temperature range to a limited extent.

The other main requirement is that the water molecules
will not fly off into space. There will always be some water
vapor in the atmosphere above the liquid water, whatever the
temperature and the pressure, and unless the gravitational
forces are sufficiently strong the velocity produced by thermal
motion will allow occasional molecules to shoot upward at
such a high speed that they will escape into space rather than
fall back again due to the pull of gravity. The escape velocity
for a rocket fired from the earth's surface is about seven miles
per second, whereas at room temperature the *average* molecular velocity for water molecules is a little greater than the
velocity of sound, about a fifth of a mile per second. This is
only an average; an appreciable fraction of the molecules of
the atmosphere will be traveling a lot faster than this, especially at higher temperatures, but the safety margin is large
enough that rather few molecules of the size of H_2O, O_2 or
N_2 are lost into space. Much lighter molecules, such as H_2,
move more rapidly, since the bigger molecules which collide

with them punch them harder because of their greater mass (H_2 has mass 2, H_2O mass 18, N_2 mass 28). Molecules of molecular or atomic hydrogen are constantly being propelled out of the atmosphere. The moon, on the other hand, though a fairly decent size, is too small for its mass to retain any of the common gases for any length of time. Any atmosphere it might have had has been lost over the many millions of years since its formation.

When considered in detail the problem of planetary atmosphere turns out to be a complicated one, depending not only on the amount and type of energy radiated from the parent star and its distance from the planet, but also on other factors, such as the amount of energy reflected by the planet's surface (which is much higher from snow or ice than from field or forest) and the amount reflected by clouds. It also depends on the molecular composition of the atmosphere. Too much CO_2 may trap the heat being radiated back by the planet, thereby causing a "greenhouse" effect. But leaving all these details aside, we can see that the minimum requirement is for a planet above a certain minimum size—a size not unlike that of the earth—at such a distance from its parent star that it is neither too hot (as Mercury is) nor too cold, as it would be if it were as far out as Jupiter is and had no additional source of heat.

There is also a restriction on the type of star. The rate at which a star uses up its nuclear fuel depends very much on its mass. A massive star consumes its fuel very rapidly. It is, therefore, very hot, radiating a lot of energy into the space around it. Any planet with liquid water on its surface must be further away from such a star than we are from the sun. This does not itself cause a problem. The difficulty lies in the relatively short time during which the star sends out light and heat. A fairly massive star may only last for as little as ten million years. This hardly seems long enough for life to evolve to any appreciable extent. The sun, on the other hand, has been radiating fairly steadily for over four billion years and is likely to do so for as long again.

Stars whose mass is appreciably less than the sun's pose a different difficulty. They can shine steadily for a much longer period, so that we need have no worries about the time available for life's evolution. Since such a star emits less energy, any suitable planet will have to be closer to it than we are to the sun. For this reason there will only be a rather small *range* of distances if the planet is to have the sort of conditions we need. A little less and the planet will be so hot that the water will boil. A little more and all the water will freeze to ice. Thus, we can expect to find some smaller stars with suitable planets, but there will be rather few of them because the exact conditions are more difficult to fulfill. Even for the stars which are the size of the sun, the range may be so small that only an occasional planetary system will have a planet in just the right place, as indeed seems to be the case for our own solar system.

In summary, we need a star which is not too big, or its lifetime will be too short, nor too small, or the chance of it having a suitable planet will be too slim. Fortunately, the sun is a fairly average star. It turns out that many stars have a fairly acceptable size. What we now need to know is whether such stars usually have planets circling around them.

There is, unfortunately, no widely accepted theory of the origin of the solar system, in spite of all the new experimental evidence which has been accumulated by space research in the last ten or more years. In the earlier part of this century it was speculated that the solar system was formed from a streamer of material dragged out of the sun by the close approach of another star. As such, it would have been a very rare event and consequently only very few stars would have planetary systems. A more detailed theoretical treatment showed that such an event is unlikely to lead to the planets as we know them. Most recent ideas are tied to the origin of the sun itself. This is believed to have been condensed by gravity from a slowly spinning cloud of dust and gas, whose rotation speeded up as the diameters of the system got smaller, due to the conservation of angular momentum. This spin produced a flattened disk of a material from which the planets were believed to have been formed due to a further condensation, again driven by gravitational attraction. Exactly how this happened—whether, for example, a nearby supernova explosion was needed to trigger the system—is not completely clear. It is thus not possible to say with complete confidence, on theoretical grounds alone, that planetary systems are likely to be common, though one might suspect this to be the case. We must, therefore, look at the experimental evidence.

This turns out to be very sparse. Planets are too small and the light they reflect from their parent star far too dim for us to be able to detect by direct observation even those revolving around the nearest stars. A large planet will influence slightly the orbit of the star around which it circles—they will both circle around a point which represents their common center of gravity. In very favorable cases it might be possible to detect the movement of such a star, and indeed in one instance it was thought that such a wobble could be detected. However, there is now more than a suspicion that the observed effects were due to experimental error, because the expected movement is so small.

At first sight, then, the problem looks hopeless. If this were the case we could only sit back and wait for novel or vastly improved methods of detection. But there is one effect which can be observed fairly easily which may give a clue. The distribution of angular momentum (roughly speaking, the amount of spin, the total of which in a closed system must remain the same) in our solar system is rather odd. Most of the spin, defined in this way, is found to be in the planets and rather little of it in the sun. It seems possible that the protosun was originally spinning much faster, with the dust cloud revolving around the sun going correspondingly more slowly. By some mechanism (and detailed suggestions have been made as to how it might have happened) spin was transferred from the sun to the dust cloud, thus slowing the former and speeding up the latter.

By good fortune the rate of spin of a star can in many cases be detected by a careful study of the light it emits, since at one of its edges the material of the spinning star may be moving toward us and at the other edge moving away from us. These movements alter the frequencies of the light that reaches us, due to the Doppler effect. It is found experimentally that stars of about the size of the sun fall roughly into two classes. Some of them spin very rapidly, as one might expect from the way they were first formed, whereas others appear to spin much more slowly. It is tempting to believe that the latter type of star has been slowed down because it has a planetary system. If this line of argument is correct, then planets will be quite common.

Unfortunately, this is really the only evidence we have for the existence of planets. One is always more comfortable in science if two or more distinct lines of reasoning lead to the same conclusion. Here we have only one. Experience has shown that such a deduction can only be accepted with reserve. Having said that, one must concede that the direct evidence for stellar rotations is really very convincing, the deduction about the existence of planets circling around slow-spinning stars fairly plausible and not incompatible with our broad theories as to how stars and planets may have arisen. On balance it seems more likely that planets are fairly common rather than very rare.

There is one other factor which we must consider about possible planetary systems. Just as it is fairly easy to detect the spin of a star from a detailed study of the light it sends us, so we can also detect double stars, that is, two stars, fairly close together, circling around one another and held in their orbits by their mutual gravitational attraction. The two stars need not be the same size nor type and in fact they are often rather different from each other. It turns out that such multiple systems are rather common, being almost the rule rather than the exception. Now, a planetary system circling around a pair of stars, each circling around each other, is likely to be rather less stable than one such as ours which has only a single star at its center. The double stars, unless they are very close together (in which case their gravitational effect in the planets approximates that of a single star), may disturb the orbits of the planets, since sometimes a planet will be closer to one star and then, a bit later, to another. This will not only make the energy falling on any particular planet vary periodically, but, more important, the planets may be in greater danger of colliding with each other. The steady conditions over long periods of time, which we think are needed for the evolution of higher forms of life, may not easily occur in such planetary systems. Thus, even though many double stars may have planets, they may not be ideal for the evolution of life. Of course, a little variation may, for all we know, be a good thing, and jerk evolution out of a rut from time to time, but it is difficult to believe any life would survive the actual collision of two planets.

There remains the problem of the planet's atmosphere.

This has already been discussed in Chapter 6. Here we must widen the discussion to embrace planets outside the solar system. As we have seen, it is difficult at the present time to decide just what the earth's early atmosphere was like. It is even more difficult when we do not know the size of the star, the exact size of the planet and how far apart they were. In the solar system Venus is not unlike Earth, being a little nearer the sun and a little smaller in size. In spite of this, its atmosphere is very different from ours, being very hot, very dense (the pressure at the surface is more than a hundred times the atmospheric pressure here) and consisting largely of CO_2. This high concentration of carbon dioxide produces a greenhouse effect, trapping the radiation attempting to flow out into space, and this, together with the greater flux of energy from the sun, raises the temperature to about $720°K$. It is this high temperature which causes so much CO_2 in the atmosphere, since it is high enough to vaporize some of the carbonate in the rocks. On Earth, although carbonate is fairly abundant—as in the white cliffs of Dover —the temperature is just sufficiently low for almost all of it to remain in solid form or dissolved in the oceans. In short, a relatively small difference in planetary conditions may make a large difference to the planet's atmosphere.

It is thus possible that a planet might be found which is more massive than the earth, though at such a distance from its star that it has liquid water on its surface. If the planet were massive enough, the abundant hydrogen in the dust cloud might be retained on the planet (as it is for our outer planets such as Jupiter), or at least lost much more slowly. The resulting atmosphere, being reducing, might be very favorable for the production of a good "tasty" soup on its surface. It is thus at least possible that there are in the universe more suitable places for life to start than any found in our own solar system.

Although the earth looks as if it were a fairly average planet circling a fairly average star, we cannot be certain that it may not have special features which made conditions especially favorable for the origin of life. A possible example of this is our moon. Moons are fairly common around the planets in the solar system, but by analogy with the other planets one might have expected the earth to have several smaller moons rather than the one big one we see shining above us. The origin of the moon has still not been determined. It seems unlikely that it was spun off from the earth. Was it formed when the earth was formed, or was it captured by the earth at a later date, having originated elsewhere in the solar system? Possibly our moon merged with an earlier set of smaller moons in the process.

Whatever the moon's origin, one would have expected it to be considerably nearer the earth in those early times. The moon raises tides on the earth. This friction not only slows the rotation of the earth, which must have been considerably faster in those days, but by a reciprocal action gradually forces the orbit of the moon to a larger radius. When the moon was closer to the earth the tides would have been bigger. Just how much bigger depends on how the moon originated and how it changed its orbit. It is possible that it may have been first captured in a reverse orbit, which then gradually contracted and changed over to the present direction of rotation, going over the poles in the process. If this were the case the tides at that time would have been very large. This may have had all sorts of effects. Without them there may have been a thick skin of hydrocarbons over all the waters of the earth. These early tides may have churned this up into an emulsion, perhaps making conditions more favorable for the emergence of primitive cells. Such large tides may have produced continual wetting and drying on a fairly large scale in pools near the margins of the oceans and seas.

Again, these may be very favorable conditions for prebiotic synthesis. In general terms, large tides would move things about and make for more variety on the surface of the primitive earth.

Another more subtle effect may have been produced by continental drift. But for plate tectonics—the movement of the various plates over the surface of the earth—there might be no mountain-building. In that case, constant weathering would erode the land, carrying the debris into the sea, as rivers do today, until at length all the land would lie beneath the oceans. This might not completely prevent the emergence of life, but if it happened early enough it might have made the origin of life more difficult. If later, higher organisms would probably have evolved very differently. It is not easy to imagine how modern science would emerge if there were no dry land at all, though it would be rash to say that it could not happen.

Mountain-building occurs probably because the interior of the earth is fairly fluid and the rocks not far from the surface sufficiently plastic to allow them to yield at an appreciable rate. These conditions are present because the interior of the earth is rather hot (a very rough estimate of this temperature makes it about as hot as the surface of the sun). This probably depends, among other things, on the radioactivity in the rocks, especially that due to isotopes of uranium, thorium and potassium. The *percentage* of radioactive atoms in the rocks is not all that high, but there is such a lot of material in the earth that in sum it amounts to an appreciable amount. Moreover, such radioactive decay produces a relatively large amount of energy. This heat, together with the heat remaining from the time of the earth's aggregation, is contained by the great thickness of the earth's crust and its low conductivity for heat, so that this small internal supply of heat can maintain a very high temperature because the insulation is so good.

With all these complications in mind, let us stand back and try to make a very rough estimate of the number of planets in the galaxy which have on their surface a watery solution of organic compounds—a thin soup from which life conceivably might emerge. The total number of stars of all kinds in our galaxy is estimated to be about 10^{11} (one hundred billion). Only a fraction of these will be the right size and only a small fraction of these will not be double stars. Perhaps one star in a hundred might fulfill these two conditions. This would allow us 10^9 possible stars. Even if only one-tenth of these had planetary systems of broadly the right kind, we should still have 10^8 of them. It is more tricky to estimate just what proportion of these would have a planet of a suitable size at just the right distance from its star, but perhaps one in a hundred might be a conservative guess. This would still leave us with a million planets in our galaxy on which we might hope to find oceans of thin organic soup waiting for life to get going.

As can be seen immediately by the crude way these estimates have been made, there is considerable room for debate concerning the most likely values for each of these figures. Our estimate of a million may be somewhat too low. The important point is that it is difficult to believe it is too high by such a large factor that there is no other planet similar to the earth anywhere in the galaxy. For this to be true our estimates would have to be at least a million times too big. Of course, we may be totally wrong about planetary systems. Conceivably they could be extremely rare—this would mean that the slow spin of stars similar to the sun had some other explanation. There may be some subtle conditions which we have overlooked, so that though planetary systems are common, the earth is really a freak and planets like it occur very seldom, if at all. Short of some direct experimental evidence, we can never be sure that our crude guesses do not have some large flaw in them. A small error will not do. Even if our estimate is one hundred times too high this would still leave ten thousand suitable planets in the galaxy. At the present time there is only one reasonable conclusion, frail though it is. Planets with a suitable soup are likely to be fairly common in the galaxy.

This does not imply that they will be rather close together. Even if there were a million of them, their average distance apart would be a few hundred light-years. If there were only ten thousand of them, their distance apart would be some ten times greater than this. Of course, our rather conservative estimates may be much too low, in which case they might be as close together as ten light-years, but such an extremely short distance seems very unlikely. Even so, it would take a rocket traveling at one-hundredth the speed of light a thousand years to travel such a distance.

NINE

Higher Civilizations

WE HAVE JUST SEEN that it seems more than likely that there exist many other planets in the galaxy having on their surfaces a large amount of a rather watery solution of organic molecules of the kind needed to act as the raw bricks with which to build a living system. We have also seen, in Chapter 6, that at the present time we can form no clear idea of whether such a soup is likely to lead to a primitive living system within a reasonable time—say a billion years—or whether most of these soups are doomed to remain lifeless almost indefinitely because the origin of life is such an exceedingly rare event. In this chapter we consider another problem.

Given that some simple replicating system had managed to get going, how likely is it to evolve to a stage in evolution similar to our own?

When we consider what we know about the stages of evolution on earth we find a rather curious thing. The simpler organisms seem to have taken the longer time to evolve. The earliest traces of life we can detect at the present time are found associated with rocks dated to about 3.6 billion years ago. Multicellular organisms probably began about 1.4 billion years ago, but the fossil record, formed by simple animals whose hard parts have been preserved, is only 0.6 billion years old. These events are marked on the time chart at the beginning of this book.

Further research may show single-celled organisms earlier than 3.6 billion years ago. Thus, the time available for pre-cellular evolution—what we have been considering the difficult step—could hardly have taken a billion years and may have taken considerably less. Against this, the time for the single-celled organisms to take the next decisive step appears to be about two billion years, or perhaps a little more. After that, evolution appears to have speeded up. The earliest mammals are only 200 million years old, and they did not really radiate to give forms similar to most of those we see today till as little as sixty million years ago.

Certainly one decisive step was the origin of the eukaryotes, those organisms with a true nucleus, a mitotic process and with mitochondria in their cytoplasm to handle their energy supply. Plants acquired chloroplasts to handle photosynthesis. One has a feeling that such a development may have been essential for the evolution of the higher animals and plants. Certainly those that did not undergo it, the bacteria and the blue-green algae, have remained relatively simple, though well adjusted to their environment.

It is unclear just how unlikely this step was. It is strongly suspected that the mitochondria in our cells are the descendants of some earlier free-living form which infected a rather different cell and then learned to live there symbiotically. Perhaps the acquisition of cell mobility and with it the ability to phagocytose—to engulf food particles and even whole organisms—made all the difference. Whatever it was, it seems to have taken a very long time to happen. This would suggest that is was a very rare event. If the probability of it happening had been, for some reason, only half its actual value it might never have taken place, even at this late time in the earth's history. The earth today might be seething with bacteria and algae, and little more.

Such an argument cannot be used for *all* the steps we see in evolution. Once primitive animals with muscles and nerves have arrived, we can be confident, given a good molecular mechanism for rapid evolution, that a visual system will develop. The ability to see gives an animal a considerable selective advantage, and, more significantly, such a development occurred at least three distinct times in evolution—in the insects, in the mollusks (such as squid and octopuses) and in the vertebrates (fish, amphibians, reptiles, birds and mammals). Anything which happens several times independently in evolution is not likely to be a very rare event. It is the steps which happened once, and especially those which appear to have taken a long time, which we might suspect to be due to a happy accident and, therefore, not to be relied on in any similar process elsewhere.

Exactly how many such steps there may have been is difficult to decide. Another possible example concerns the extinction of the dinosaurs. About sixty million years ago the dinosaurs, which at that time were the dominant vertebrates, especially on land, suddenly became extinct, together with a large number of other species of both animals and plants. Two physicists, Alvarez and Alvarez (father and son), and their colleagues, noticing that there was a thin layer of clay deposited at about that time, analyzed it and found that it had a peculiar isotopic composition, containing an unusual amount of iridium. Three widely separated locations were examined and all three had this clay layer, suggesting that it was produced by some worldwide event. The isotopic composition was compatible with an extraterrestrial origin for some of the material. They have proposed that an asteroid, some six miles in diameter, hit the earth, producing a tremendous cavity and scattering a large amount of material into the atmosphere which, spread by winds all over the world, blocked out the sunlight for several years until at length even the finest dust particles had time to settle. (It is still remembered that after the explosion of Krakatoa there were striking sunsets all over the earth for several years, because of the dust suspended in the atmosphere.) As a result of the virtual extinction of daylight, many plants would have died, especially the phytoplankten in the ocean. Many species would have become extinct, though plants with longer-lasting seeds could spring up again when the light eventually returned. As a result of this massive loss of plant material the food chain was totally disrupted. This would have been especially lethal to the larger animals at the top of the food chain. Thus, all the dinosaurs became extinct except possibly a few small ones, the ancestors of the birds. The earliest mammals had evolved about 200 million years ago, but at the time of the catastrophe they had not had a great deal of success, probably because they were kept down by the dominant dinosaurs. These early mammals were mainly small, nocturnal insectivores and might thus have survived the years of darkness. When the light eventually returned, the mammals rapidly evolved to occupy all the various ecological niches vacated by the now-extinct dinosaurs (just as Darwin's finches radiated on the Galapagos Islands), soon forming the many species whose descendants we see all around us today. One branch, the primates, developed good color vision and an enlarged cerebral cortex, eventually producing man.

The key question is whether the dinosaurs, if they had been left undisturbed, would have evolved any animal intelligent enough to develop science and technology. This we cannot answer with any certainty, but one has a sneaking suspicion that the dinosaurs had specialized in the wrong direction. If so, then the evolution of a higher intelligence on earth depended crucially on this very drastic jolt given to evolution by the asteroid. Such a collision may not have been a unique event. There are other, earlier extinctions in the fossil record. One might expect a large asteroid to hit the earth with an average frequency of about one in 200 million years, though it has not yet been documented whether these earlier extinctions were due to such an impact.

It is possible that evolution, in the long run, will always produce a creature with a high degree of intelligence, because in the struggle for existence, intelligence usually pays. But it may need rather large changes in the environment to bring to fruition the larger steps in evolution. If so, this places another condition on planetary systems likely to evolve higher forms of life within a reasonable time.

TEN

How Early Could Life
Have Started?

So FAR we have considered *where* life might have arisen in the universe and how rare an event it is likely to have been. We have not considered at all *when* it might have started, nor how long a time is required to progress from the earlier beginnings to a higher civilization capable of sending rockets to other planetary systems. Strictly speaking, we can form no firmer estimate about the time needed for evolution than we can for the chance of any particular step, except that one would be reluctant to believe that the whole process could have happened very much more quickly than it did on earth. There is no detailed theory of evolution so quantitative that we can calculate just how long any particular stage is likely to require. We can see that it may depend on factors such as the rate of mutation, the generation time, the size of the interbreeding population and above all on the selective pressures produced by the environment in general and other species in particular. Environmental stability, a large interbreeding population and long generation times will tend to produce slow changes. Isolation mechanisms, geographical or biological, leading to small populations may produce new species rather rapidly. Any species which finds itself with many opportunities and few rivals, as often happens when new lands are first colonized, is likely to diversify very rapidly to fill all the new ecological niches available. But in all these cases it is difficult to *predict* the rate of evolution, except in a very rough and ready way. In a very deep sense evolution is a process whose course is *necessarily* unpredictable. Only when a particular attribute is likely to provide an overwhelming advantage (such as being able to see) can we be fairly confident that it is bound to emerge, one way or another. Even in that case we would be rash to predict exactly what form a visual system would take. The most we can say of the nervous system of a higher animal is that it is likely to evolve so that the animal perceives and responds not merely to the obvious signals falling on its sense organs but to those features of these signals which correspond to particular aspects of the real world, and especially those aspects which will affect the animal's survival and reproduction—the smell of a predator, the appearance of the female and so forth. But how long it will take an animal's brain to evolve a particular sophisticated function is almost impossible to answer exactly.

The easiest question we could attempt to answer might be this: if life on earth started all over again, with only trivial changes in the environment (so that it would not repeat the process exactly), how long would it take for a creature like man to arise? We know that this process originally took about four billion years. It could conceivably happen again in as little as a billion years, but a time much less than that strains our credulity. On the other hand, a much longer period might be required if one or two happy accidents were missing. It seems almost impossible to decide, and for life on a rather different planet the difficulty is compounded.

I shall thus be compelled to break the rules imposed by the theory of probability and assume that life, once having started elsewhere, would evolve at about the same rate as it did here; that is, it would take roughly four billion years from soup to man. From everything one has said one can see that such an assumption is hopelessly insecure unless one can show that *all* the major steps in evolution had a fairly high probability. If this were true then chance delays and chance accelerations might tend to average out and the overall rate might be much as it was here. Even this assumes that general factors, such as temperature or the multiplicity of environments available, were not so different that the whole course of evolution, though broadly similar to that on earth, was either appreciably speeded up or slowed down. All one can really say is that the figure of four billion years for the whole process, though lacking any firm support, is not outrageous.

Armed with this very dubious number, we can start to consider when life might first have arisen. There are two essential requirements. We need a suitable planet and we need certain elements in or near its surface. Clearly we cannot have these too soon after the Big Bang, as discussed in Chapter 2. There is good evidence that many of the atoms in our bodies were not formed during the early moments of cosmic expansion but were synthesized by some of the first stars. These large stars used up their nuclear fuel rapidly, collapsed, exploded and scattered their debris into the surrounding space, whence it was eventually condensed to form new stars and planetary systems. Though we cannot be sure just how long a time was needed for all this to happen on an appreciable scale, a reasonable estimate might be one or two billion years.

To proceed any further we need to know the age of the universe—the time since the Big Bang. Unfortunately, this is still a matter of controversy. The higher estimates range up to twenty billion years, the smallest as little as seven billion, though few people would accept such a low figure. When Leslie Orgel and I wrote our paper the best guess seemed to be about thirteen billion. Today a figure of ten billion might be thought nearer the mark.

For our purpose the exact figure is not essential, provided it does not turn out to be too short. To be on the safe side, let us take ten billion years. Allowing one billion for the evolution of planets and chemicals, this leaves nine billion years. We see immediately that this is about twice the age of the earth. There is enough time for life to have evolved not just once, but *two times in succession.* In short, the time available would allow for life to have started on some distant planet formed nine billion years ago, for creatures like ourselves to have developed four to five billion years later and for them to have then sent some simple form of life to the earth, which by that time had cooled to the stage when the primitive oceans had already formed. Whether this really happened is as yet a matter of opinion, but on the present evidence it is difficult to argue that the total time available was certainly too short. There was ample time for life to have evolved not just once but twice over.

ELEVEN

What Would They Have Sent?

FROM THIS POINT ON we must leave behind quantitative considerations, however approximate, and allow our imagination a somewhat freer hand. We shall postulate that on some distant planet, some four billion or so years ago, there had evolved a form of higher creature who, like ourselves, had discovered science and technology, developing them far beyond anything we have accomplished, since they would have had plenty of time and it is most unlikely that their society would have stopped at exactly the stage at which we are now. Just how much further they would have got it is not easy for us to guess, though some of their science may have been not unlike ours. Our knowledge of many parts of physics and chemistry is now so complete and on such solid foundations that their main features may already be known to us. This is unlikely to be true for *all* parts of these subjects. High-energy physics, for example, probably still holds many surprises in store. We can expect new methods in

117

physical chemistry which will make our knowledge of chemical structure and chemical reactions more exact. Even if no radically new principles remain to be discovered (and this is rather unlikely), there is work for generations of scientists, discovering in detail how atoms and molecules interact in many different mixtures and in many conditions of pressure and temperature.

When we turn to astronomy, astrophysics and cosmology we realize that in these fields much more remains to be discovered. We have already touched on some of these problems—how many stars have planets, for example—and there are also unanswered questions on a grand scale, such as whether the universe is opened or closed (that is, whether it has enough mass so that eventually it will fall back on itself, rather than go on expanding forever). Our knowledge of biology is even more primitive—we still have only the sketchiest ideas about the details of embryology, for example, and, as we have seen, the course and mechanism of evolution are still only understood in outline and the origin of life even less so.

We can be confident that if our own civilization survives for as little as a further one thousand years we shall have answered many of these difficult questions. Even if all the fundamental principles of all the sciences are established by then, much will still remain to be done. Within the next ten thousand years we can expect many complex systems to be worked out in fair detail. Above all, we are likely to see an enormous flowering of engineering projects, applying the fundamental knowledge then known to systems of ever-increasing power, subtlety and complexity. Provided mankind neither blows itself up nor completely fouls up the environment and is not overrun by rabid antiscientific fanatics, we can expect to see major efforts to improve the nature of man himself. What form these may take, how successful they will be and how much time will be needed to change human nature radically we can hardly surmise as we peer through the fog of uncertainty which envelops the distant future.

By analogy we may expect that these earlier technocrats were likely to have known much more than we do, especially in astronomy and biology, and to have developed a technology far in advance of ours. How will their universe have looked to them?

It would be surprising if they had not fathomed the secrets of their own nature (which we are very far from doing), the mechanisms of their evolution and the detailed workings of their immediate physical surroundings. Whereas we can only guess which stars may have planets, they are likely to have known, though how much they would have known of the exact conditions of these other worlds it is difficult to estimate. Given a higher technology and enough time, we might expect them to have sent unmanned space probes to at least a few of the nearest stars and, after a delay of some hundreds of years, to have received messages back conveying something of the conditions there. Even to do this would have required a technology far more developed than ours.

Let us assume that they had discovered that there are many places in the galaxy suitable for life, possessing both land and oceans, with a steady supply of radiation from their parent star, a suitable atmosphere and, in consequence, very large volumes of diluted soup on their surface. What we cannot guess so easily is whether they were able to discover how many places had some primitive form of life. Perhaps they found that life is indeed a very rare event. Even if the converse is true, it is possible that they may have erroneously concluded that they were effectively unique and that no other forms of life existed in their galaxy. We can imagine, without straining credulity too far, that as they looked around their little corner of the universe, stretching for some tens of thousands of light-years in many directions, they might have concluded that while soups were common, life was extremely rare; that many places had the potential for life but in not one of them had the vital first step been taken—the spontaneous occurrence of the chemical mechanism needed for natural selection. And so, we must ask, if that was indeed how the universe appeared to them, what would they have done?

There is one further factor needed to delineate more precisely their predicament. They would have known that in the long run—and it may have been a very long run—their own civilization was doomed. Of course, there may have been reasons for them to believe they could not even survive in the short run. Perhaps they had found that a neighboring star was set on a collision course with theirs—not a very likely event in most parts of a galaxy but more than likely near the galactic center. Perhaps they had reason to suspect that their social system would not be stable idefinitely, as indeed ours may not be. But they would have known that in the very long run—meaning within billions of years—when its nuclear fuel began to run out their star would probably become a red giant and in doing so would engulf their planet and roast them beyond all reasonable hope of escape. Without doubt they would have planned to colonize neighboring planets, but this may have proved to be a technological achievement of extreme difficulty, especially if they were unlucky and the nearest suitable planet was many tens of light-years away. Even if they attempted to do this, they may have realized that their chances of success were small and that they had to make contingency plans against repeated failures of this kind. Whatever their reasons, we may expect them to have examined carefully other alternatives.

What other options would have been open to them? The easiest would have been to send unmanned space probes, but these could not easily be made to reproduce, in spite of what science fiction writers believe. Not only would it have been difficult to build a machine for which the raw material would be conveniently available, but the problem of making such machines work reliably with little help from the home base, especially after the long journey in space and the trauma of landing on a distant planet, are formidable. There would have to have been elaborate mechanisms for self-repair and these, too, would be liable to an appreciable failure rate. The only favorable circumstance would have been that serious competition would have almost certainly been absent. There would have been no moths to corrupt nor thieves to break through and steal. Only the slow ravages of rust and other chemical and mechanical forms of decay would have to have been coped with.

There remained the obvious possibility of sending some other living creatures from their planet. Though necessarily lower in the evolutionary scale, the hope would have been that they might survive and multiply, and, with luck, evolve eventually into a higher form of life. If it was too difficult to send manlike creatures on that appalling journey, why not try to send mice?

Unfortunately, the advantages of using mice are rather slight. A mouse occupies less space than a man, but it has nothing like the same control over its environment. Its preservation, even as a breeding colony, over hundreds of years in a spaceship, presents very considerable difficulties, even allowing for many ingenious forms of recycling. The environments it is likely to find on arrival are almost certain to be uniformly hostile. In particular, they would be expected to lack oxygen, an almost fatal handicap in the long run. Obviously we need an organism which can be sent in fairly large numbers, which could survive the long journey in space fairly well and which would have some chance of surviving both the act of delivery onto the surface of the planet and the environmental conditions it would find there. Put this way, we see that microorganisms similar to our bacteria would have been a good choice to be the colonists sent to start life in a distant place.

What are bacteria like? The main division of the biological kingdom is not, as one might be tempted to believe, into animals and plants. Nor is it between organisms with only one cell and organisms like ourselves with many cells. The most significant division is between organisms whose cells have a nucleus like ours, called *eukaryotes,* and humbler organisms which lack such a nucleus, known as *prokaryotes.* The term "higher organisms" as often used by biologists can be very misleading. Surely we are higher organisms and so, broadly speaking, are the sort of animals you see in a zoo. But to a biologist a yeast cell, such as those which ferment beer and wine and are used to leaven bread, can be described as a higher organism. In this terminology "lower organisms" means the prokaryotes. This term comprises all the bacteria, of which there are a great many different types, and what used to be called the blue-green algae. The other types of algae are eukaryotes, as are amebae, ciliates and many other small unicellular creatures.

The division of the biological world into these two very broad categories is important because it is both clear and profound. It is not just a matter of the cell nucleus but involves many features of the internal architecture of the cell. These could not have been studied effectively without the use of modern equipment, such as the electron microscope which allows us to visualize the components of cells in much finer detail than was ever possible before. For this reason the eukaryote-prokaryote classification is comparatively recent, dating only from about 1960.

What is the difference between them? In broad terms, the eukaryotes have highly developed chromosomes which after replication are partitioned in a process known as *mitosis,* which requires a special mitotic apparatus. The "chromosomes" of prokaryotes are much simpler and they lack the molecules to make a mitotic spindle. Eukaryotes have many special components in their cytoplasm, including complicated membrane systems (which prokaryotes usually lack) and special little organelles, such as mitochondria. These have their own DNA and their own apparatus for protein synthesis and are widely believed to have descended from a free-living prokaryote which entered the cell and eventually degenerated so that it could only exist symbiotically within the host cell. A mitochondrion is commonly called "the powerhouse of the cell," since it contains the molecular apparatus for the efficient combustion of food using molecular oxygen. One of our own cells contains hundreds, if not thousands, of them.

Perhaps the more significant difference between eukaryotes and prokaryotes concerns how substances can get in and out of the cell. In eukaryotes there are special mechanisms to engulf larger particles—a process known as *phagocytosis*—and special internal structures to digest them. Prokaryotes totally lack these molecular mechanisms. Only objects of molecular size can transverse their membranes.

We do not need to know all the details. In broad terms the prokaryotes are simpler, lacking the special molecules which permit the more sophisticated eukaryotes to carry out elaborate processes. These processes have allowed the eukaryotes to have much more genetic information (by permitting a set of chromosomes instead of only one piece of DNA), to live on other organisms and to move molecules around inside themselves in a purposeful manner. If there is one property which sets the eukaryotes above the prokaryotes, it is the molecular apparatus for generating and controlling movement within the cell. This is what has led to the formation of muscles, essential for animals, and permits the complicated dance of the chromosomes we see as mitosis.

Why, then, if bacteria are so disadvantaged, should we consider them as possible passengers in our rocket? The key to the answer is one word: oxygen. It is fairly likely that in the prebiotic world there was rather little oxygen in the atmosphere. Consequently we must examine the organisms we have on the earth today to see what their oxygen requirements are.

The great advantage of oxygen is that it permits a cell to obtain far more energy from metabolizing its food. This process is usually called *respiration.* A few bacteria can use certain inorganic compounds, such as carbonates, nitrates or sulphates, instead of oxygen, but these compounds are the very ones which were unlikely to be found in any quantity on the primitive earth, because of the lack of oxygen in the atmosphere. Without an inorganic electron acceptor, as such compounds are called, the cell must use the much less efficient process called *fermentation.* The importance of fermentation is that it can proceed in the total absence of oxygen, but it produces far fewer molecules of ATP, the energy currency of the cell, then does respiration.

Molecular oxygen is a powerful but dangerous compound. It is potentially a highly toxic substance for cells, because cellular processes are liable to produce several lethal derivatives of it, such as hydrogen peroxide (H_2O_2) or an even more dangerous compound, the free-radical superoxide (O_2-).

Many cells have special enzymes to mop up these life-threatening substances. Some species of bacteria lack such enzymes. To them oxygen is a poison and they can only live in places where there is none, such as deep mud, but on the primitive earth they would have been at no particular disadvantage.

Oxygen—which one should recall is produced nowadays as a byproduct of photosynthesis—is so useful to most organisms that they cannot live without it any more than we can. It is for this reason that we need not consider most of the more highly evolved cells as candidates for space colonization. This requirement eliminates all contenders except certain bacteria and a few very simple organisms, such as yeast. Some of these can use oxygen if it is available—again, yeast is an example—whereas others cannot use oxygen at all. Some of the anerobes can tolerate oxygen, but others are killed by it.

After this preamble let us see what bacteria are like. There are so many different types that any brief description must necessarily be rather sketchy. They are usually rather small, which perhaps is not too surprising since they have only a modest amount of DNA, in the range of a million base-pairs. A typical dimension, though there is a large range of sizes, would be about one or a few microns (a micron is a thousandth of a millimeter), so they are usually little bigger than the wavelengths of visible light, which range around half a micron. For this reason, though they can be seen under a high-powered light microscope so that their approximate size and shape (whether spherical, rodlike or strung together into long chains) can be observed, other techniques are needed to reveal their secrets. Fortunately, certain bacteria have proved ideal for modern biochemical methods so that an enormous amount of work has been done on them, especially in the last thirty or forty years. This has revealed them to be truly remarkable creatures.

It might be thought that being so small, they would lack chemical versatility, but this is very far from being the case. Many of them can live on a very simple chemical medium, containing little more than a source of carbon, a source of nitrogen (such as ammonium, NH_4^+) and some compound, usually but not always organic, they can use to obtain energy. Many of them do not need most of the vitamins, since they can synthesize them for themselves rather than obtain them from their food, as we do. Nor do they need the "essential" amino acids which we get from breaking down the protein in our food, since these, too, they can manufacture for themselves. Many of them are mobile. They can move around, using their rather simple flagella, and can detect the concentration of food molecules and, by a simple strategy, can swim in that general direction. In a similar way they can avoid certain toxic substances. Under favorable circumstances they can grow and divide very rapidly. In a rich broth with plenty of oxygen they can divide into two in as short a time as twenty minutes. In a less favorable circumstance they may take half a day to double, but even at that rate they have the potential for increasing in numbers dramatically, provided the food supply lasts. They have efficient mechanisms for controlling their metabolic machinery, so that enzymes which are temporarily not needed (because of a rich food supply) are turned off and no longer produced until the cell senses a need for them again. Metabolically they seem to be streamlined for rapid growth, since in many circumstances it is the fastest cells which will win and, by natural selection, produce the subsequent generations. They have only a very tenuous sex life. Most of the time one cell simply divides into two daughter cells without any sexual process, but occasionally, by a special mechanism, two bacteria can conjugate, one (the "male") passing some of its DNA to the other (the "female"). This process can be relatively slow, the act of transfer taking as much as two hours, or several normal life-times.

Because sexual reproduction is not essential, a bacterial colony can be grown from a single individual. Moreover, since they are not compelled to search out a mate to reproduce they can grow very far apart from one another.

Bacteria usually have rigid cell walls outside the delicate plasma membrane which is the effective barrier at the molecular level between the inside of the cell and the outside. This wall prevents the plasma membrane from damage and especially from the osmotic swelling which would otherwise be produced if the cell found itself in too watery a solution. Thus, many bacteria are not too fussy about the concentration of salts and organic compounds in the medium surrounding them. Another advantage is that they can usually be "freeze-dried," a process which first cools the bacteria and then extracts the water from them in such a way as to produce minimum damage to the cellular structures.

Bacteria on earth are of many different types and live in many different environments, from hot springs to barren deserts. Some have even evolved so that they can thrive under conditions of intense radiation, such as is found in nuclear reactors. Others can utilize unusual compounds, such as hydrogen sulphide (H_2S), ferrous ions or methane, though they usually need oxygen to do this. If they also are capable of photosynthesis they could probably manage without oxygen. Other bacteria are strictly anerobic and can use hydrogen, producing methane in the process. Others can "fix" nitrogen; that is, they can obtain their supply of nitrogen from the very inert N_2 molecule in the atmosphere. Still others can carry out various types of photosynthesis, getting their energy from sunlight. It would lead us into too many technicalities to discuss all the possibilities.

There is one group of microorganisms which should be

given a little more attention here. This is the blue-green algae, or the blue-green bacteria, as they are now called, if only because the earliest known microbial fossils seem to be of this type. The group as a whole is very various, though its members do have some characteristics in common. They can all obtain energy from light. Some of them can also grow in the dark, though rather slowly, and they can only use a rather limited range of carbon compounds for this purpose. Rather strikingly, many of them can also fix nitrogen. If so, they need very little to live on, since they can grow in a medium having only a few salts, using the light to obtain carbon from CO_2 and nitrogen from N_2. Such organisms usually consist of chains of cells joined end to end, the nitrogen fixation being done mainly by special cells, called heterocysts, which are specialized to do this and which never divide again.

Not surprisingly, the blue-green bacteria live in a great variety of habitats, being found not only in the sea but in fresh water and in the soil. Some thrive in hot springs, others in deserts, where they inhabit crevices in the rocks.

After this very sketchy description of the bacterial world, let us now recapitulate some of the advantages these little creatures have for space travel. As we have seen, many of them are rather small. A fairly typical bacterium, such as *Escherichia coli,* is about one micron wide and two microns long. Thus, a billion of them can be packed into a volume of a few cubic centimeters. They can be frozen alive and most of them will survive when they are eventually unfrozen. In this frozen state they can persist almost indefinitely without any serious loss. At a very low temperature, such as that of space, many of them might well survive for well over ten thousand years. They would be almost immune to impact shock and other similar hazards. Best of all, if they fell into a prebiotic ocean they would probably thrive, especially since many species can survive with little or no oxygen. In fact, some bacteria can grow on such a simple medium that almost any prebiotic soup would allow them to survive and multiply rather effectively, provided it was not too cool. Moreover, they do not need to be huddled together. A single bacterium could, under favorable circumstances, infect a whole ocean.

Bacteria may be extremely simple when compared to organisms like ourselves, but as self-reproducing chemical factories they are not only compact and robust, but chemically very versatile. As far as I know, no one, rather surprisingly, has deliberately tried to grow bacteria in an artificial "soup" made in a Urey-Miller type of experiment (most of the experimenters go to great lengths to exclude microorganisms from their incubation flasks), but one would certainly expect many types of bacteria to thrive there, even in the absence of atmospheric oxygen.

For all these reasons, then, microorganisms, and especially those that can live without oxygen, are the obvious creatures to send to another planet, provided one's aim is to get life started there rather than to deliver a fully formed higher organism having some chance of survival. This is why Orgel and I suggested them as the most likely cargo for the unmanned spaceship we postulated for Directed Panspermia.

TWELVE

The Design of the Rocket

BEFORE WE SEE how a rocket might be designed to send microorganisms to another planet let us consider first how to send astronauts. To propel it at high speeds, such a spaceship will need a very powerful rocket motor and a good supply of fuel. It must have room both for the astronauts and also for their life support (such as food, oxygen, etc.) on the long, dark journey, together with all the instruments for monitoring and controlling it and for communicating with the parent planet. There should be enough fuel left to decelerate the spaceship on arrival and to land the astronauts safely on some planet or asteroid of the chosen star. Neither the acceleration nor the deceleration must be so fierce as to damage the passengers. The spaceship need not bring them back, since they are colonists rather than travelers.

For obvious reasons it would be an advantage if they went fairly fast. If they could travel extremely fast, close to the velocity of light, relativistic time dilation would operate.

131

While the journey, as viewed from either star, might require thousands of years, within the spaceship only a few tens of years might have elapsed. This is one of the most surprising conclusions that can be drawn from the Special Theory of Relativity.

It turns out that for people (as opposed to electrons), time dilation is almost impossible to achieve, not merely because of the very advanced technology needed but because of the basic laws of physics dealing with energy, power and mass. For example, the rocket will need a lot of energy but must not be too heavy. Thus, a very energy-rich fuel is required. We know of no better way to produce this than through the process of annihilation of antimatter by matter, but the problem of storing the antimatter safely seems insuperable. The next-best method is to use nuclear fusion and by such means turn hydrogen into helium, ejecting the helium backward to provide the thrust. Edward Purcell has calculated that even with an "ideal" rocket engine of this type the speed of the exhaust would only be one-eighth that of light. In practice it would be less than this. A rocket becomes inefficient as it reaches speeds much greater than the exhaust speed. Such calculations show that the mass of the rocket plus fuel would have to be immensely greater than the payload in order to reach speeds close to that of light.

Apart from the problem of accelerating the spaceship to these very high velocities, to say nothing of decelerating it on arrival, another major difficulty is the protection of the spaceship from damage. Much of space is quite empty, but there are occasional atoms and molecules there, and even tiny pieces of dust. Though these may themselves be traveling quite slowly, they will hit the spaceship very hard, due to the spaceship's own speed. At moderate speeds the spaceship's cargo and machinery can be protected by a thick layer of material acting as a shield. At very high speeds—approaching that of light—the thickness required becomes impossibly large.

Various ingenious suggestions have been made to overcome these difficulties. Instead of having to carry a lot of fuel, the spaceship might scoop up matter in space and use that as fuel. Even if this could be done successfully, the matter there is so extremely sparse that the collector would have to be enormous—as much as a hundred miles in diameter. Perhaps damage could be avoided by deflecting the matter to one side, but this, too, seems a heroic undertaking. The only ideas which seem remotely viable are based on providing the energy for propulsion not in the spaceship but from the parent planet, by a laser beam, for example. This should allow the spaceship to be relatively small and light (though still a substantial size), since it would not have to carry the immense amount of fuel needed by more conventional methods. Even these techniques, which are very far in the future, are unlikely to propel the spaceship faster than half the velocity of light. At such a speed, time dilation is a comparatively small effect. Tentatively we can conclude that relativistic space travel is impossible.

This means that the time the crew will experience will be simply the distance of the journey divided by the average velocity of the spaceship. To travel a hundred light-years at one-hundredth the speed of light would take them ten thousand years. For all but the shortest journeys in very advanced spaceships the voyage is likely to last longer than the human lifespan (of course, creatures which have evolved elsewhere may live longer). Either life must be prolonged in some way —by freezing the astronauts, should this prove possible—or the astronauts must breed in the spaceship; not my idea of the Good Life.

Enough has been said to show that sending space colonists with any hope of success is a tremendously difficult undertaking, far in advance of our own present powers. The ingenuity, persistence and effort required is so enormous that I myself rather doubt that we or even our descendants will accomplish it successfully, though we cannot know what the future has in store.

After such a series of way-out suggestions it comes as a relief to turn to the more mundane problem of sending simple bacteria to another planet. To do this I shall only consider techniques which, though impossible now, are not too far in the future.

Having decided that the spaceship might go fairly fast, though not as fast as light, it is not easy to guess its most likely velocity. We can ourselves build spaceships which would leave the solar system with speeds in the range of 3 miles per second, that is, .0015 percent of the speed of light. Without going into the details—whether nuclear explosion could be used, whether a beam from earth could supply the energy for acceleration and so on—it seems fairly certain that a spaceship could be designed which would travel at one-thousandth the speed of light. To raise the speed to one-tenth the speed of light looks rather difficult. A reasonable guess for the velocity might be one-hundredth that of light.

Within one hundred light-years of the earth there are several thousand stars and, because of the arguments given earlier, it would not be too surprising if one of them had a planet with an environment of the kind our bacteria would need. Naturally, at an earlier stage in the life of the universe the stars may have been more widely separated. Alternatively, this earlier civilization might have emerged in a part of the galaxy where stars were appreciably nearer together. However, the chance of a suitable planet being within ten light-years seems rather small, within a thousand light-years more than likely, so that one hundred light-years is probably as good a guess as any.

This gives the time for the journey as, very roughly, ten thousand years. By our everyday standards this is an immense time, but we must ask whether it is so long that the expedition would have been certain to fail. Nobody has stored bacteria in the cold for anything approaching such a time, but the results we do have for much shorter periods suggest that if they are frozen carefully and kept cold enough, many bacteria will survive such treatment for very long times indeed. It seems highly likely that further research would easily yield ways to preserve bacteria for periods as long as ten thousand years, and perhaps for as long as a million. In any case, such large numbers of them can be carried that quite substantial losses can be tolerated, provided at least a few remain to colonize the new environment.

A more serious problem would be to make sure that the spaceship worked reliably after ten thousand years in space. This is because the rocket must function not only at the start of the journey but also at its termination. The delivery of the bacteria to the planet is not a straightforward matter. It is not practicable to shoot the rocket into empty space and hope for the best. Stars are so sparse that at any reasonable speed the spaceship would probably go right through the galaxy and come out the other side. A suitable star would have to be selected as a target and the rocket kept on course during its journey. This would be a relatively easy matter. The main problem arises when the spaceship finally approaches the star. At this point it must decelerate, which implies that it must have carried rocket fuel all that way and that the rocket motors and control system must still be in good working order. Then the spaceship must be able to select a suitable

planet, home on that and then release its cargo in such a way
that it will survive the entry through the atmosphere and
splash unharmed into the primitive ocean. None of this
seems overwhelmingly difficult, but it demands a very highly
developed technology in order that the various rocket components will work reliably after the extended journey in
space. The problems look soluble, but in the long run rather
than the short run.

Whatever the details of the spaceship, it seems likely that
it could have carried and delivered very many microorganisms. Judging by our own rockets, a pay-load of two hundred pounds would not have been unreasonable. Bacteria
are so small that 10^{16} to 10^{17} might be stored in such a space.
Because this number is so large the bacteria could have been
packaged in many separate parcels. This would have made
delivery very much easier. During delivery these packets
could have been scattered throughout the atmosphere, so that
they could land in many different places on the planet's
surface. Each packet would have to have been housed in a
casing which could withstand both the frictional heat generated as it passed through the atmosphere at high speed and
the shock of impact as it hit the ocean (those that hit land
may well have been lost). Once in the water the coating
would have had to dissolve, thus releasing the bacteria. All
these requirements look as if they could be easily satisfied,
given a little ingenuity. Multiple delivery has the advantage
that even if many of the packets fall in unsuitable places, a
few of them may be lucky and find a propitious environment.
Very few bacteria may be needed to infect a sterile planet—
even one might be enough, provided it could grow and divide successfully.

Since many bacteria would have been sent, it would have
made good sense to send more than one variety. Exactly how

these would have been chosen is difficult to judge, since it would have depended to some extent on what microorganisms were available on the planet from which the rocket was sent. Since the atmosphere of the new planet is unlikely to have had much oxygen, it would have seemed rather a waste of time sending microorganisms which preferred to metabolize their food by using oxygen. It would have seemed better to send those which were preadapted to whatever conditions might have been expected on the new planet. All of them might have used organic compounds as an energy source, but others might have been able to use the energy stored in certain minerals. Photosynthesis would have seemed very desirable, and perhaps the ability to form spores, at least for some of the organisms. The senders could well have developed wholly new strains of microorganisms, specially designed to cope with prebiotic conditions, though whether it would have been better to try to combine all the desirable properties within one single type of organism or to send many different organisms it not completely clear. Whatever the best solution, it does not seem to present very serious difficulties and, in fact, such a research program could be carried out today, since we are now beginning to develop very powerful methods for modifying the genetic composition of organisms, and especially microorganisms. A study in 1976 on the habitability of Mars concluded that the best type of microorganism might be based on our present blue-green algae. As stated earlier, it is striking that the earliest known microfossils on earth appear to be of exactly this type.

It is more difficult to decide how advanced a microorganism would have been sent. If the overwhelming requirement was to start any form of life, however simple, and if this was thought to be a hazardous and difficult undertaking, then the simpler and more rugged the microorganism, the better.

If it was thought that once having reached a suitable planet, the initiation of life there would be relatively easy, then it would have made good sense to send some more advanced microorganisms, to give evolution as much of a head start as possible. If we ourselves were trying to select microorganisms, we would certainly be tempted to try to send some form of eukaryote—that is, a cell with chromosomes, a true nucleus and useful macromolecules, such as actin and tubulin which help to give mobility to both the cell and its components. Yeast would be an example of such an advanced cell. It thrives on oxygen but it can also live without it.

If such organisms were sent to the earth at the beginning of life here, we can see little trace of them in the fossil record. The present eukaryotes, as far as we can judge, arrived on the scene a lot later. It could always be argued that some type of eukaryote was originally sent here, but that it failed to compete with the better-adjusted bacteria, perhaps when the original supply of food in the primitive oceans was used up, and so it died out. Alternatively, it may have discarded many of its fancier attributes and have evolved into something simpler and more able to cope with the struggle for survival. If a mixture of microorganisms were sent, it would be surprising if, once established, life then died out altogether, so tough and versatile are these tiny creatures, but without experiment one would hesitate to predict just what type of organisms would come out on top in an environment so remote from our everyday experience.

From this discussion, one thing emerges very clearly. In the environment of a prebiotic ocean, especially beneath a nonoxidizing atmosphere, certain microorganisms have an immense advantage over any higher forms of life. As described in the previous chapter, they are chemically versatile, oxygen is not essential to them and, being small, they can multiply very fast. Add to this their very desirable qualities as passengers—their small size, their ability to stand freezing and thawing, their comparative lack of sensitivity to radiation damage—and we see that they are almost ideal for interplanetary fertilization. It may be true that a man will, in time, be able to travel a certain distance in space beyond the narrow confines of the solar system, but whatever that distance is, *bacteria could go further*. And, as far as we can see, this advantage is likely to remain, however big the advances in technology.

This point becomes important should one wish to answer those who are convinced that space travel will eventually be relatively easy, since they would argue that if men could be sent, it is superfluous to bother with bacteria. Should this turn out to be correct, there is still one hypothetical situation in which Directed Panspermia would be an advantage. Suppose an advanced form of life developed four billion years ago in a neighboring galaxy, such as Andromeda, but was entirely absent from our own. Though these versatile creatures may have succeeded in colonizing the whole of Andromeda, the technical problem of hopping to a neighboring galaxy may have been too great even for them to tackle. Realizing that they themselves could never travel the million or so light-years in space from Andromeda to our galaxy, they saw, as we do, that bacteria could go further, and sent off spaceships filled with microorganisms. Though it is not easy to see how to make a spaceship suitable for such an immense journey, it would be rash to say that it is impossible, since it is very difficult to foresee all the technical advances which the future may bring.

Since bacteria are such ideal passengers, is there any form of propulsion which would work for them but not for men? There is at least one. A good example of a radically different approach to the rocket problem is the suggestion by Mautner and Matloft that advanced solar sails might be used to power the spaceship. Such sails would have to have a large area and be extremely thin, so that the pressure of the sun's radiation would exceed the attraction of the sun's gravity. The authors estimate that sails with a mass of about a tenth of a milligram per square centimeter (and such materials are already available) would be thin enough to allow the spaceship to escape from the sun. Even thinner sails would make it leave faster. By these means it is difficult to reach very high speeds, such as one-hundredth the speed of light (0.01C), but speeds between one-ten-thousandth and one-thousandth (0.0001 to 0.001C) could probably be achieved. These relatively low speeds would limit the range of the spaceship somewhat, since even at a speed of .001C it would take ten thousand years to go as small a distance as ten light-years. This is rather restrictive, but it must be set against the immense advantage of the proposal, which is that the deceleration required at the end of the journey could also be done by the solar sails, so that no large supply of fuel is needed for this operation, although a very small amount would be required to propel the many little packets of bacteria in the payload into orbits where at least some of them might be captured by the waiting planet.

The authors estimate that for a payload of about ten tons the sails might have to have a radius of about 200 yards. The details of such a spaceship are very different from the more conventional ones, but they support once more the proposition that *bacteria can go further.* This is likely to be true whatever the method of propulsion and whatever the range of the spaceship, whether it be a mere ten light-years by solar sails or the long journey of two million light-years from Andromeda by some very much more advanced technology.

THIRTEEN

The Two Theories Contrasted

THE PRECEDING ARGUMENTS all sustain the thesis that Directed Panspermia is not implausible. This means that we have two types of theory about the origin of life on earth and that they are radically different. The first—the orthodox theory—states that life as we know it started here all on its own, with little or no assistance from anything outside our solar system. The second—Directed Panspermia—postulates that the roots of our form of life go back to another place in the universe, almost certainly another planet; that it had reached a very advanced form there before anything much had started here; and that life here was seeded by microorganisms sent on some form of spaceship by an advanced civilization.

The two theories could hardly be more different, but it is important to ask, does the difference matter? Since the universe in its present form had an origin in time—the Big Bang—and since any form of life at such early times was impossible, life must have started somewhere at some time well after the Big Bang. It could be argued that Directed Panspermia merely transfers the problem elsewhere. This is partly true, but for all we know the location was vital. It may emerge eventually that for one reason or another it would have been almost impossible to start life on Earth, whereas on some more favorable planet it could begin more easily and perhaps evolve more rapidly. Perhaps our unusual moon will turn out to be more of a handicap than an advantage. Thus, although we cannot as yet give any powerful reasons why an origin elsewhere was much more plausible, it is rash to assume that conditions here were just as good as anywhere else. Whether life originated here or elsewhere is, at bottom, an historical fact, and we are not entitled, at this stage, to brush it aside as irrelevant.

The two theories, then, are radically different. Can we decide which is more likely to be correct? In particular, can we marshal any convincing evidence which might support or refute Directed Panspermia? One possible line of evidence is contained within the organisms we have today. In spite of the great variety of molecules and chemical reactions produced by evolution, there are certain features which appear common to all living things. As we painstakingly collect more and more data from organisms alive today, we can begin to piece together the family trees of certain molecules —transfer RNA molecules, for example—in the hope of being able to deduce the nature of the earliest ancestors of these molecules. Such work is still in progress, but there is one feature which is so invariant that it immediately attracts attention. This is the genetic code, described in the Appendix. With the exception of mitochondria, the code is identical in all living things so far examined, and even for mitochondria the differences are rather small. This would not be surprising if there were an obvious structural reason for the details of the code; if certain amino acids had necessarily to go with certain codons because, for example, their shapes fitted neatly together. Brave attempts have been made to suggest how this could happen, but they all seem unconvincing. It is at least as plausible that the details of the code are mainly accidental. Even if certain early codons were not dictated by chance but had some chemical logic to them, and even if some broad features of the code can be explained in some way, it seems most improbable, at least at the moment, that all the *details* of the code were decided by purely chemical reasons. What the code suggests is that life, at some stage, went through at least one bottleneck, a small interbreeding population from which all subsequent life has descended.

Now, there is no strong reason why such a bottleneck should not have occurred during the earlier stages of evolution on the earth. One version of the code may have been so much better than any other, may have given its possessors such a selective advantage over all its competitors, that it alone survived, all the others becoming extinct. Nevertheless, one is mildly surprised that several versions of the code did not emerge, and the fact that the mitochondrial codes are slightly different from the rest supports this. However, out of the many different types of organisms on earth rather few have been deliberately checked to determine their exact genetic code. Since it is suspected that the code will always be the same, few people are keen to spend time on the problem. Perhaps, with further work, more varieties will be found. Till this happens the fact that the code is so uniform lends a small measure of support for Directed Panspermia.

Is there any other feature, common to all living things, which appears unusual? In our original paper Orgel and I suggested that the element molybdenum appeared to be more abundant in living things than one might have expected from its natural abundance in the rocks. Several people pointed out that while molybdenum was rather rare in rocks, it was much more common in sea water. To this Orgel replied that while this was true of today's oceans, it seemed unlikely that molybdenum was present in such amounts in the prebiotic ocean, since the greater reducing conditions at that time might have made its salts rather soluble. Even if Orgel's argument is accepted it must be conceded that the support it gives to Directed Panspermia is rather feeble. Even if there was rather little molybdenum in the prebiotic ocean, the early organisms may have learned to concentrate it within themselves in some way.

Perhaps a better approach might be to ask what special features we might hope to see in the fossil record if Directed Panspermia had indeed occurred. The main difference would be that microorganisms should appear here suddenly, without any evidence for prebiotic systems or *very* primitive organisms. We might also expect that not one but several types of microorganisms would appear which, although distantly related, would be rather distinct. In particular, it might be difficult to trace intermediate ancestral forms, since these would only have existed on the sender planet, not on Earth. Of these distinct forms we should not be surprised to find one which resembled the blue-green algae, since this has independently been suggested as a good candidate for an effective primitive organism.

Now, it is perhaps remarkable that these are all features of the early fossil record or of the early evolutionary trees deduced from the study of present-day molecules. The earlier fossils, so far, do indeed resemble the blue-green algae. They date to a relatively early time in the life of the earth, so early that one is surprised to find them fully formed at that stage. Attempts to trace back molecular family trees seem, at the present time, to lead back to several distinct families which appear rather distant from each other. Thus, at the very least one can say that this evidence does not contradict Directed Panspermia but supports it to some extent.

Unfortunately, a more careful examination of the evidence reveals that this support is rather weak. We do not have available a whole series of sedimentary rocks dating back from 3.6 billion to 4.6 billion years before the present era, or thereabouts. Thus, it is not surprising that we lack evidence for earlier forms. We may be struck by how soon the blue-green algae arose in evolution, but they would have had about a billion years to do this, and since we have no way of calculating the rate of prebiotic evolution by any independent method, our "surprise" at their appearance at that time merely reflects our ignorance combined with our earlier expectation (for no good reason) that microorganisms appeared later. The molecular family trees, though suggestive, are at the moment too fragmentary to lend any strong support to any theory. Once again, we can only say that these data do not contradict Directed Panspermia, though they might be considered suggestive.

We must then look to the other side of the debate. Are there good reasons for rejecting Directed Panspermia? Certainly there are one or two lines of argument which might make one uncomfortable.

One of these concerns the age of those stars which contain a reasonable abundance of heavy elements. Their age must be several billion years less than the age of the universe. At the moment the latter number is still a matter for debate. If further work supports a figure at the shorter end of the range, then the age of the most suitable stars might be as little as six or seven billion years. This would leave rather a short time for the origin and development of the postulated higher civilization which sent out the rocket, perhaps as little as two or three billion years. On reflection, we see that this argument does not have much force. Why should not two billion years be enough? We have seen that the longer phase of evolution on earth was that solely occupied by microorganisms, a period of two billion years or more. If on the other planet this phase was shortened to, say, half a billion years, and if the prebiotic phase was not too long, then it does not seem impossible for a higher form of life to have evolved from scratch in two billion years. Put another way, if the latter stages of evolution on earth—that revealed by the conventional fossil record, from the earliest creatures with hard parts right up to man—took only 0.6 billion years, why should not the earlier stages, in perhaps more favorable circumstances, have gone just as rapidly? It is thus difficult to refute Directed Panspermia for this reason unless it can be shown that the sun is, in reality, one of the oldest stars of the type required. On the present evidence this seems unlikely.

Perhaps the most telling argument against Directed Panspermia is the lack of any sign of eukaryotes in the earlier rocks. If we ourselves were sending microorganisms to a distant planet we would certainly try to dispatch one or two carefully chosen eukaryotes in company with several of the more obvious prokaryotes, carefully chosen because all the many species we have on earth can metabolize their food using oxygen, a much more efficient process than glycolysis, the method by which food is handled without oxygen. However, only a minority of earthly eukaryotes can exist without oxygen, yeast being the prime example. It would, therefore, seem sensible for us to develop special eukaryotes, derived from the ones we have here, which were specially designed to

live under prebiotic conditions, since even if we sent something like present-day yeast it would seem likely that it
would soon lose the capacity to use oxygen in an environment
that had little or none. Unfortunately, the same tendency to
lose potentially useful attributes may apply to other eukar-
yotic specialties. For example, it has been argued that the
basic reason for the success of the eukaryotes, and their ability
to radiate into many different species, was their ability to
phagocytose—to eat other, usually smaller creatures. This
made a food chain possible, and with it the opportunity for
considerably more diversification. To do this, eukaryotes
have evolved several unique molecular structures, microtu-
bules, actin, myosin and so on, which help them to move
and to engulf other creatures. But in prebiotic conditions,
especially after infection by Directed Panspermia, the ocean
was unlikely to be teeming with microorganisms, since there
was probably not enough food to sustain a dense population.
On the contrary, one might expect that at these earlier stages
cells were few and far between. In such circumstances an
organism which was potentially capable of eating others
might come across too few of them to make them anything
but a rather minor source of food. Natural selection might
well have caused the organism to discard all these rather
superfluous molecular structures, which would have cost it
energy to make, and forced it to concentrate instead on evolving those which could make better use of the soup. The one
other property which would have been of considerable value
would have been photosynthesis, and we would certainly
have dispatched some organisms which could carry out this
complex but very rewarding operation, since the more energy
a cell could get from the sun, the less it would need to obtain
from the soup. But the earliest fossil cells we have appear to
be of just this type, the blue-green algae. Again it seems as
if the argument against Directed Panspermia has rather little

force and the evidence appears, if anything, to support the idea, though very weakly.

We are, therefore, in a very unsatisfactory position. We have two distinct theories, very different from each other, and yet we seem unable to estimate which one of these is the more likely to be correct, let alone decide decisively between them. Why is this? Are the theories deficient in some way, or is the subject one of special difficulty?

There seem to me to be two different criticisms of Directed Panspermia, which are diametrically opposite in character. The first, which my wife has voiced more than once, is that it is not a real theory but merely science fiction. This is not meant as a compliment, though it might perhaps be taken that way. There is a story that an intelligence agency once assembled a collection of rather distinguished scientists without explicitly saying why it wanted their advice. At the beginning of the meeting the agency explained that it had decided that it needed to know what scientific advances were coming in the future, so that it could be prepared for the possible impact of the resulting technology on the various tasks the agency had to perform. At this, a well-known physicist rose and said that the wrong set of people had been invited. "We are all too sound," he said, "and this makes us conservative. The people you should consult are the science fiction writers. *They* are the ones who can see, far more clearly than we can, what the future holds in store."

There is some truth in this, though it does require a little sifting of the wheat from the chaff. Early science fiction writers, such as H. G. Wells and Jules Verne, have a rather presentable record, describing men on the moon, remarkable submarines and so forth. The converse is also true. Leading scientists have made a series of foolish remarks about what will *not* happen. But this was not what my wife had in mind.

What she was implying was that the idea had too many of the trappings of conventional fiction—the superior civilization elsewhere, the rocket with exceptional powers (a phallic symbol?), even the busy little microorganisms infesting the virgin earth. How *could* such stuff be considered seriously? The whole idea stinks of UFOs or the Chariot of the Gods or other common forms of contemporary silliness.

Against this I can only claim that whereas the idea has indeed many of the stigmata of science fiction, its body is a lot more solid. It does not really have the major feature of most science fiction, which is a great leap of the imagination, glossing over the wildly improbable scientific foundations from which this leap is made. Each of the details which contribute to the required scenario are based on a fairly solid foundation of contemporary science: the age of the universe, the likelihood of planets, the composition of the prebiotic ocean, the toughness of bacteria in adversity and the ease with which they can flourish where most other organisms would surely die, the design of the rocket and so on. The whole idea is, in fact, rather unimaginative; it might be described as a tissue of plausibility.

And this leads us to the other criticism, that the idea is in fact too pedestrian to be true, that it needs only our present technology plus the sort of logical development of it which a few tens of years will bring. Yet, such a critic would say, this supposed advanced civilization, if it ever reached the level we have today, would surely have gone on much, much further, to attain levels of science and technology which we cannot even glimpse. So it is not foolish to conduct the argument using only what we know today as a basis? Will this not inevitably turn out to be false?

There is some force in this argument, but several rebuttals could be attempted. In the first place, I would contend that

Orgel and I were trying to construct a scientific theory, and it is not scientific to wave one's hands about and proclaim that in the long run all things are possible. Moreover, we do not, in fact, have at this moment the technology to send bacteria to any planet outside the solar system, though, as we have argued, we have a good basis for this technology. Nor is the idea of Directed Panspermia necessarily restricted to the rather straightforward realization which we have sketched. New technologies could make for better possibilities and at least for a greater chance of success than anything we could hope for in this century at least. Finally, if forced into a corner, I would resolutely raise the banner "Bacteria can go further" and claim, though not without some qualms about what the future might bring, that *whatever* new technology were invented, this slogan would still be true. There will always be a range beyond which the only practical objects to send are bacteria. To those who might say that in the centuries ahead projects of this type would be all too easy, I would ask, "Could your rocket go to Andromeda? And if it could, what would you send?"

All these arguments seem to me not very fruitful because they do not seem to be getting at the heart of the matter. What we should be concentrating on is not so much the flavor of the idea but its status as a scientific theory in good standing. When we do this we see that there are indeed other deficiencies.

The first is the nature of the evidence used to construct the theory. Much of this is little more than painting in the background. There is just one piece of evidence which might give serious pause for thought and that is the apparent universality of the genetic code, though, as we have seen, this is far from being solidly established.

The snag is that it was mainly because we were brooding on this rather odd fact that Orgel and I hit upon the idea of

Directed Panspermia. This means that according to the rules —at least the rules that I play by—it should be given little or no weight in the *testing* of the theory. The hallmark of a successful theory is that it predicts correctly facts that were not known when the theory was presented, or, better still, which were then known incorrectly. A good theory should have at least two characteristics: it should be in sharp contrast to at least one alternative idea and it should make predictions which are testable. A third desirable property, that it should be a deep theory—that is, it should apply over a very wide range of observations—is not really applicable here.

Directed Panspermia certainly fulfills the first requirement. It is when we come to the second one that we get into trouble. The theory makes a fairly strong prediction: that the earliest organisms should appear suddenly, without any sign of more simple precursors here on earth. A second prediction is plausible but not essential for the success of the theory: that several distinct types of microorganisms should appear more or less simultaneously. Clearly, if we had a full fossil record of the earliest cells we should be able to settle the matter one way or the other, so the theory is not completely vacuous.

The essential difficulty, then, is not so much the nature of the theory but the extreme paucity of the relevant evidence. Not only are there few sedimentary rocks from that epoch which have not been mangled inside the earth at some step during their long history in the earth's crust, but even if we had a good selection of them (and it seems likely that, in time, rather more will be found than we have now) it would still be difficult to obtain enough to be certain that vital evidence is not missing. When we consider how difficult it has been to track down in detail the evolutionary history of even such a large animal as early man, especially when we take into account how recent that evolution was on a geological time scale, we see that the task of tracing the evolution of the earliest cells on earth is formidable. Neither do the other alternatives look very promising. The hope that living things contain "molecular fossils" in certain of their macromolecules is a valid one, but something very striking would have to emerge to permit us to decide decisively between the two theories. The same is true for simulated prebiotic experiments. It is true that two lines of evidence, because of their dramatic nature, have given us hope. The complementary nature of the structure of DNA and RNA on the one hand, and the Miller-Urey experiment on the other, are both so striking that it would be surprising if they were not relevant to the origin of life. But will there be other such experiments? Can protein synthesis be carried out today in the test tube without any ribosomes, using only messenger RNA and some prototransfer RNA loaded with amino acids? If this worked it would indeed be dramatic. Could we have a really convincing prebiotic synthesis of RNA from elementary components, which produced sufficiently long chains with a reasonable degree of accuracy? And even if we could do all this, would it really make the origin of life here so overwhelmingly certain that the idea of Directed Panspermia would seem superfluous?

In deciding between two theories, one soon learns that plausibility alone will not do, quite apart from the fact that it is usually contaminated with our unstated prejudices. Directed Panspermia may at first sight seem farfetched, but can we give solid reasons for this initial reaction? Thirty years of experience in molecular biology has taught one the lesson that plausibility is not enough. It will not do just to put the nail on end and give it a little tap. It is essential to drive it home. To give a theory the degree of certainty we need, we have to hit it hard, again and again. And this, alas, is just what we are unable to do in this particular case. Every time I write a paper on the origins of life I swear I will never write another one, because there is too much speculation running after too few facts, though I must confess that in spite of this, the subject is so fascinating that I never seem to stick to my resolve.

The kindest thing to state about Directed Panspermia, then, is to concede that it is indeed a valid scientific theory, but that as a theory it is premature. This inevitably leads to the question, will its time ever come? And here we must tread cautiously. The history of science shows that it is all too easy to state, with the best of scientific reasons, that such and such will never be discovered or that so and so will never be done. "We shall never know of what the stars are made." "Nuclear energy will never come." "Space travel is bunk." What is striking is how short a time it took to upset these negative prophecies. It is not that I believe that all things are possible. I would instance levitation as something I think is wholly unlikely.* But leaving levitation aside (it is, incidentally, a good test to sort out the scientifically-minded from the simple-minded), it is too easy to make rash negative predictions. I cannot myself see just how we shall ever decide how life originated, but I believe that at least the evidence on which to base such a decision will grow, though when, if ever, it will reach such a level that we can feel confident that we have found the answer, only the future can tell. All we can say is that the problem, and the related problem of life on other worlds, is so important to us that in the long run, it will be bad luck if we fail to find the answer.

* By levitation I mean sustaining oneself by an act of will for a minute or so in the air, well above the ground, without help and *with no gadgets* (as opposed to learning to jump into the air from a sitting position on a mattress, so that one imagines oneself to be levitating).

FOURTEEN

Fermi's Question Reconsidered

Now that we have Directed Panspermia in perspective we should return very briefly to Fermi's question: if there are intelligent beings elsewhere in the galaxy, why are they not here?

Michael Hart has argued that since there are no signs of them, this must imply that we are the only form of highly evolved life in our galaxy. His main point is that if they exist at all it is unreasonable to imagine them frozen at exactly our stage of development and, therefore, they are likely to have produced a very advanced form of technology, which, he believes, would enable them to manufacture spaceships capable of traveling distances of tens of light-years, at speeds such as one-hundredth to one-tenth that of light, and there found new colonies. After these colonies had had time to consolidate and expand in their new homes they in turn would send out spaceships to found further colonies. In this way they would hop their way from planet to planet till they had spread all over the galaxy.

How long this would take depends on a number of factors —the speed of the spaceships, the average consolidation time, whether they expanded always outward, or moved in successive journeys in a more random manner, and so forth. The surprising result is that however the rate is calculated, the time to cover the whole galaxy is not as great as might have been expected: perhaps less than a million years, though some combinations of figures give times up to 100 million years. Bearing in mind how much earlier life might have started elsewhere, because our planet was born relatively late, Hart argues that they should have reached Earth by now.

As the reader can see from the discussion earlier in this book, this argument is very far from being watertight. It may turn out to be a very demanding undertaking to construct spaceships which can carry passengers to other suitable planets and which will allow them to set up a colony there, so demanding that some of these higher civilizations may never have constructed them. Perhaps they tired of technology before then and adopted a different lifestyle, either lapsing into idle pleasure, as Gunther Stent has predicted we shall do, or cultivating a purely spiritual way of life, possibly supported by specially designed psychedelic drugs. Perhaps they destroyed themselves, as many fear we shall do, by their advanced nuclear technology. This is especially likely to apply to those cultures aggressive enough to want to venture into space. Even if all higher civilizations did not lapse in one of these ways, so that there were always some which succeeded in building suitable spaceships, the wastage may have been not inconsiderable.

If and when they did get to a new planet, its environment was probably so unfavorable that they would have had to modify it extensively to make it suitable for their habitation. If they needed oxygen, as seems very probable, they may have had to cultivate agriculture on their new home on an extensive scale so that plants could produce the oxygen for them. They may even have had to carry out extensive genetic engineering in their spaceship before this afforestation could be done successfully, because the peculiarities of the atmosphere and the rocks of the planet may not have been suitable for the plants they brought with them. This environmental modification could have taken so long that there may have been a real danger that, due to some accident, the whole colony would become extinct. After all, not all the early American colonies were successful; several were abandoned for one reason or another. Even if they had finally succeeded in founding a new civilization, their descendants might have preferred to live there for a very long time before venturing once again to undertake the difficult and traumatic project of further colonization.

For all these reasons there may have been so much wastage that the process was not a continuing one. For life to spread indefinitely, each civilization must, on an average, send out so many colonies that at least one will survive to send out, in time, a similar number of colonies. In short, there may have been a few attempts to spread all over the whole galaxy, but we cannot be sure that they did not all fizzle out after the first few steps.

On the other hand, if they had decided, perhaps only as an interim measure, to try Directed Panspermia and had sent microorganisms, they could have constructed such spaceships fairly early in their technological development, before they either destroyed themselves or lost interest, and the spaceships could have had a much larger range. Against this they would have realized that the consolidation time might now be billions of years, compared with the mere thousand or ten thousand years needed to expand a colony of spacemen to occupy a whole planet. Of course, they might have considered Directed Panspermia a useful long-term way of producing an oxygen atmosphere in many places which their remote descendants could one day use to their advantage.

Even if Hart's mechanism of galactic colonization is accepted, people have been reluctant to agree with his conclusion that we are alone in the galaxy. Instead, they have preferred to believe that a higher civilization may indeed have spread all over the galaxy, but for one of several reasons these colonists are not obviously apparent to us today. Very few astronomers are prepared to give any credence to UFO sightings, if only because the percentage of obviously false reports is so high. It is true that there is always a residue of unexplained observations, but it is not encouraging to learn that when UFO reports increase, due to scares or to promotion by the media, the number of unexplained reports also increases, suggesting that they, too, are probably without significance.

However, there is nothing to rule out the possibility that Earth was inspected transiently at some time in the past, say forty million years ago, and then abandoned as unsuitable. Perhaps the visitors did not feel that our planet would make an ideal home for them, or perhaps they were ecologically-minded and did not wish to disturb the local flora and fauna. John Ball has suggested that we may be part of a cosmic wildlife park, which is being left so it can develop undisturbed. Perhaps we are under some sort of discreet surveillance by higher beings on a planet of some nearby star. It is not clear exactly how these cosmic game-wardens would do this without our detecting them, but with a higher technology such supervision may be relatively easy. In any case, we are now betraying our presence by our TV programs which, as microwave noise, are escaping into space and spreading outward with the velocity of light.

Another suggestion is that they may have arrived in the solar system but have not chosen to visit Earth. Michael Papagiannis has proposed that they may be living on their spaceship in the asteroid belt, using the sun's light as a source of energy and the asteroids as a source of raw material for their industrial operations. Against this, David Stephenson has remarked that he would have expected them to be further out, skulking near the orbit of Neptune and only from time to time making forays to the asteroids to obtain carbonaceous material.* It would be extremely difficult for us to detect even a very large spaceship as far away as Neptune's orbit or even in the asteroid belt, since the asteroids would tend to camouflage it. None of these suggestions can be ruled out as impossible, but they sound too much like science fiction—the assumptions are too extreme and the chains of reasoning too long. One is reluctant to embrace them without at least some other type of evidence in their favor.

Of course if, as I have argued, it may be very difficult for life to get going, Hart's conclusion that we are alone in the galaxy may be correct even though his reasons for it are, as we have seen, not especially convincing.

If there are indeed other intelligent beings in the galaxy and if for some reason or another they have stayed at home, perhaps they are trying to send us signals of some sort. This is too complicated a topic to develop fully here. Signals are far, far easier to send than rockets, but even with them there are difficulties. What wavelength should be used? Should the radiation be sent in all directions, or transmitted in a narrow beam to make it go further? If so, in which direction should it be sent? *What* should be sent? The prime numbers are a

---

* Papers by Papagiannis and Stephenson are contained in Donald Goldsmith's collection, *The Quest for Extraterrestrial Life. A Book of Readings* (see Further Reading).

favorite introductory signal, because they are the same every-where, as is much of mathematics, physics and chemistry. Unfortunately, liberal arts subjects, such as literature and history, are likely to be almost unintelligible to another civilization, at least at first contact. Whether their music would in any way resemble ours is debatable.

Even if many civilizations exist in the galaxy it is not a foregone conclusion that they would be transmitting signals into space. Should we ourselves send messages? As Tommy Gold has said, if we'd any sense we'd keep quiet. Perhaps everybody is listening and nobody is talking. There have been modest attempts to listen for such signals, both in the U.S.A. and in the U.S.S.R., so far without success. Whatever one feels about the probability of other life in the galaxy, an inexpensive program to listen for possible signals would seem to make very good sense, especially as it may lead to useful astronomical knowledge as a by-product.

In the long run Fermi's question demands an answer. Once the scale and nature of the galaxy is appreciated it is intolerable not to know whether we are its sole inhabitants. It may even be very dangerous not to do so. If the discussions in this book show anything, it is that deciding the matter one way or the other will not be easy. It remains an outstanding challenge for our science and our technology, both for us and for our descendants.

FIFTEEN

Why Should We Care?

AT THIS POINT the reader may feel slightly cheated. If life started so long ago, and if it is so extremely difficult to discover just how it happened, why should we bother about it? Ordinary men and women, going about their daily affairs, might well claim that, whatever the outcome, it would make no difference to them.

I believe this view to be misguided for two reasons, a particular one and a general one. To permit me to make my point, let us suppose that starting a self-replicating system of the right type from primitive earth components is not, as we have feared, an almost impossible task, but is instead a relatively easy one. It might turn out that by an ingenious choice of components and conditions a living system might form itself in the laboratory within a comparatively short time, such as a year or even less. I find it difficult to believe that such a discovery would not have a tremendous impact on almost every educated person, and especially on the younger ones. The psychological effect of being able to demonstrate in a graphic way something actually happening can be very great, as can be seen by the remarkable impact on people's view of their own planet produced by the photographs of the earth taken from space, though I doubt whether any laboratory experiment could have as great an aesthetic appeal as the pictures of our beautiful cloud-patterned globe hanging in space.

Reproducible experiments demonstrating that a rudimentary living system can evolve from a purely chemical nonliving one should strengthen our feeling of unity with nature in the broadest sense, meaning with the atoms and molecules of which all materials on the earth are made. Whether such a discovery will also have the "practical" consequences so beloved by senators and business people (what practical value does a ballgame have?), by which they mean: can it be used to cure something? will it make money? I really do not know, though few fundamental scientific discoveries have lacked some form of useful application.

But, a critical reader might reply, you are surely not entitled to use this argument, since it is just as probable, if not more probable, that we shall not be able to produce such an experimental demonstration in the foreseeable future. The actual chemical origin of life may be an extremely rare event and too elusive to reproduce here and now, especially considering the rather modest scientific effort being put into the problem today.

To such an argument I can only offer a rebuttal in very general terms. I would base my position on the very remarkable situation in which the human race finds itself after five or ten thousand years of civilization. The Western culture in which most living scientists were raised was originally based on a well-constructed set of religious and philosophical beliefs. Among these we may include the idea that the earth was the center of the universe and that the time since the creation was relatively short; the belief in an irreducible distinction between soul and matter; and the likelihood, if not certainty, of a life after death. These were combined with an excessive reliance on the alleged doctrines of certain historical figures, such as Moses, Jesus Christ and Muhammad.

Now, the remarkable thing about Western civilization, looked at in the broad sense, is that while the residue of many of these beliefs are still held by many people, most modern scientists do not subscribe to any of them. Instead, they have a quite different set of ideas underlying their view of life: the exact nature of matter and light and the laws which they obey; the size and general nature of the universe; the reality of evolution and the importance of natural selection; the chemical basis of life and in particular the nature of the genetic material; and many others. Some of these theories have the names of scientific "prophets" associated with them, such as Newton, Darwin and Einstein. These men are held in high regard, yet their ideas are not regarded as beyond criticism, nor are their lives considered to be especially praiseworthy; it is their works that are valued.

A modern scientist, if he is perceptive enough, often has the strange feeling that he must be living in another culture. He knows so much and yet he is acutely aware of how much remains to be discovered. He feels keenly that we need to understand these profound mysteries and also that with time, effort and imagination we can do so. This gives a great feeling of urgency to his quest, especially as he is not ready to accept uncritically traditional answers which lack scientific support.

While there is little active hostility to his point of view —creationists are a nuisance, but so far only a minor one—

he is nevertheless puzzled by the response to his work. A considerable fraction of the public shows a keen interest in the discoveries of modern science, so that he is frequently requested to give lectures, write articles, appear on TV and so on. Yet even among those who are interested in science —and many people are indifferent or somewhat hostile—it seems to make very little difference to their general view of life. Either they cling to outmoded religious beliefs, putting science into a totally distinct compartment of their minds, or they absorb the science superficially and happily combine it with very doubtful ideas, such as extrasensory perception, fortune-telling and communication with the dead. The remark, "Scientists don't know everything," usually identifies such persons. Scientists are painfully aware that they don't know everything, but they think they can often recognize nonsense when they come across it.

It is only in the last ten years that people have recognized many implications of the idea that man is a biological animal who has evolved largely by natural selection. Even now very few professors of ethics approach their subject from this point of view. Hardly anyone, observing the massive attachment of the public to organized sport, asks himself why so many people behave in this very strange way, and even fewer wonder whether the widespread enthusiasm for football is perhaps partly a result of the many generations our ancestors spent in tribal warfare.

The plain fact is that the myths of yesterday, which our forebears regarded not as myths but as the living truth, have collapsed, and while we are uncertain whether we can successfully use any of the remaining fragments, they are too rickety to stand as an organized interlocking body of beliefs. Yet most of the general public seems blissfully unaware of all this, as can be seen by the enthusiastic welcome given to the Pope wherever he travels.

Of course, many modern philosophers have accepted this general position, but the majority of them seem so devastated by the collapse of the old beliefs that they exude nothing but a rather dismal pessimism. Only scientists seem to have grasped the nettle. This is mainly because they are buoyed up by the tremendous success of science, especially in the last hundred years. While a scientist is sobered by the economic and political problems he sees all around him, he is possessed of an almost boundless optimism concerning his ability to forge a wholly new set of beliefs, solidly based on both theory and experiment, by a careful study of the world surrounding him and, ultimately, of himself and other human beings. Only someone actively groping with the intricacies of the brain can realize just how far we have to go in some of these problems, but even in that case the feeling is that within a few generations we shall have got to the heart of the matter.

It is against this background that we must approach the origin of life. We then see that it is indeed one of the great mysteries which confront us as we try to discover just how the universe is constructed and, in particular, to locate our own place in it. It ranks with the other major questions, many of them first clearly formulated by the Greeks: the nature of matter and light, the origin of the universe, the origin of man and the nature of consciousness and the "soul." To show no interest in these topics is to be truly uneducated, especially as we now have a very real hope of answering them in ways which would have been regarded as miraculous even as recently as Shakespeare's time.

The origin of life is also closely related to one further major question which has only been touched on in this book: are we alone in the universe? To discuss this here in detail would take us too far afield, since many other aspects of the problem have to be considered—exactly how to send and receive signals over vast distances, for example. It is mainly because we cannot estimate whether the origin of life is very rare or very common that we have yet to come to close grips with the question of other intelligent beings in space. Notice that if an earlier civilization sent microorganisms here in a rocket, it is more than likely that they dispatched many rockets, some to stars near them other than the sun. This might imply that even if life in the galaxy is exceedingly rare, there may nevertheless be several other planets which became infected some four or so billion years ago. Such planets are now likely to be very distant from us, due to the dispersive motion of stars as they rotate slowly around the galactic center, so that even if life in such places has by now reached an advanced state of development it may be too far away for us to communicate with it easily at the present time.

One does not need much imagination to realize the sensation that would be caused if an authentic message were received from another civilization. It is only because this eventuality seems so remote that people do not stay awake at night worrying about it. Our descendants may take a different view as, with highly sophisticated instruments, they peer into space, trying to describe what is there, whether there are signs of life in any form and, above all, how to explore this vast and empty universe we see all around us.

EPILOGUE

Should We Infect the Galaxy?

ONE TOPIC REMAINS. Even if it turns out that we shall never know for sure how life began here, we may still be confronted, at some time in the future, with the practical question: should we attempt to start our form of life elsewhere in the universe? And if so, how should we do it?

Many of these issues have already been discussed in Chapter 8. We may expect that by that time (if we have not destroyed ourselves by our own folly) we shall be able to decide whether the nearer stars have planets, perhaps by placing sophisticated new instruments on the moon. We may even know, more or less, how our solar system was formed, thanks to extended exploration of other planets, the asteroid belt, comets and so on. This may enable us to estimate which planets are likely to possess a fairly favorable environment. The design of rockets may be expected to have improved enormously, so that they can go very long distances and work reliably for very long times even if they cannot approach the speed of light.

With all this at our disposal, what should we do? Perhaps one of the easier things would be to try what was done for Mars; not to send men, at least in the first instance, but to send instruments which could report back to us. Even this apparently simple requirement seems technologically far in advance of what we can do today. It would require difficult feats of engineering to get a spaceship successfully into orbit, especially after such a long journey and at such a distance. In orbit it would sense far less than if it could settle on the solid surface of the planet (if it had one), yet to get it to the surface would require an even more advanced technology. Some of these problems might be solved if people were sent on the mission, but this poses a whole new set of problems, not the least being how to make sure that they arrive alive. The chances of their starting a colony, under the very unfavorable conditions likely to be found there, or of making the return journey alive seem infinitely remote. Ironically, as Tommy Gold has suggested, the most likely outcome, if they got there at all, would be that some of the bacteria they carried would reach the primitive ocean, there to survive and multiply long after the death of the astronauts. In that case, why not just send bacteria in the first place? Immediately we decide on this option, all our design problems become simpler, as we have already described in earlier chapters. If there is one intellectual exercise which disposes the mind to look more favorably on Directed Panspermia, it is that of imagining what we ourselves might do in the future exploration and colonization of space.

But notice that in our enthusiasm for infecting our neighbors there is one little detail that we have overlooked. What if our chosen planet has already evolved another form of life? Whether our descendants will have been able to decide that life was very common in the universe or, alternatively, very rare, we cannot know. We cannot even estimate how good their guesses might be. The technology to decide whether a nearby star has planets and, broadly, what they are like does not seem too far distant, but the technology needed to decide whether they possess life or not, in one form or another, would seem to be very far in the future. We can see these problems on a smaller scale as we try to discover whether there is some form of life on the planets and moons of our own solar system. The only good evidence comes from bodies on which a landing has been made. The urge to explore space is likely to reach a high level long before we can know whether what we shall be exploring supports any form of life.

It is difficult to see what would be the outcome of such a situation. Our descendants will be confronted with novel problems in cosmic ethics. Are we, as highly developed beings, entitled to disturb the fragile ecology of another planet? Should we feel bound to respect life, *whatever* form it may take? We have similar dilemmas on earth, as any vegetarian will tell you, though not many people would respect the smallpox viruses' right to life. Perhaps there will be a profound division of opinion among our descendants, though I cannot help thinking that it will be the meat-eaters who will want to explore space and the vegetarians who are likely to oppose it.

I should say, in passing, that I do not think these fears apply to the spaceships which we are at present sending outside the solar system. Even if they harbor any bacteria at all, those few microorganisms are highly unlikely to survive both the journey in space and the entry into another solar system. The chance of their infecting another planet is so extremely low that we would be foolish to worry about it.

One obvious plea is that the matter should not be rushed. Since, with luck, we have millennia ahead of us, and since as time goes on we should know more and be able to tackle more difficult tasks, why hurry? But even this argument assumes that the world will be politically stable for an indefinite period. If it is not, there would certainly be pressure from powerful groups who wanted to get on with the job, lest circumstances arose in which it could never be completed. My prejudice would be not to press ahead too eagerly, if waiting is at all possible. We should not lightly contaminate the galaxy.

APPENDIX

The Genetic Code

THE GENETIC CODE is the small dictionary which relates the four-letter language of the nucleic acids to the twenty-letter language of the proteins. Each triplet of bases corresponds to a particular amino acid, except for three triplets which signal the termination of the polypeptide chain. The code is set out in a standard form, reproduced opposite, which, because it uses abbreviations, takes a moment or two to understand. The four bases of messenger RNA are represented by their first letters: *U*racil, *C*ytosine, *A*denine, *G*uanine. Each of the twenty amino acids is represented by three letters, usually the first three letters of its name. Thus GLY stands for *GLY*-cine, PHE for *PHE*nylalanine.

As an example, consider the top left-hand corner of the code. We see that both UUU and UUC code for phenylalanine, since PHE is written in that position. In the bottom right-hand corner, we find that glycine (GLY) is coded by all four triplets beginning with GG, that is GGU, GGC, GGA and GGG. Most amino acids have several "codons," as they are called, but tryptophan has only one—UGG—as has methionine—AUG.

Rather surprisingly, the triplet AUG is also part of the signal for "begin chain," since all chains start with methionine or a close relative. This initial amino acid is usually clipped off before the protein is completed.

The code given opposite is the standard code, used by the vast majority of protein-synthesizing systems found in animals, plants and microorganisms. This chart does not reflect the fact that some small modifications have been recently found. According to this new information, the genes inside human mitochondria use both UGA and UGG for tryptophan. AUA codes for methionine rather than isoleucine. Thus, in human mitochondria all amino acids are coded by at least two triplets. There are four STOP codons instead of the usual three (UGA is now tryptophan) since AGA and AGG also code for STOP rather than for arginine.

Other species of mitochondria, such as those in yeast, are similar though the deviations from the standard code are not exactly the same as those for human mitochondria.

THE GENETIC CODE

1st ↓ 2nd →	U	C	A	G	↓ 3rd
U	PHE	SER	TYR	CYS	U
	PHE	SER	TYR	CYS	C
	LEU	SER	STOP	STOP	A
	LEU	SER	STOP	TRP	G
C	LEU	PRO	HIS	ARG	U
	LEU	PRO	HIS	ARG	C
	LEU	PRO	GLN	ARG	A
	LEU	PRO	GLN	ARG	G
A	ILEU	THR	ASN	SER	U
	ILEU	THR	ASN	SER	C
	ILEU	THR	LYS	ARG	A
	MET	THR	LYS	ARG	G
G	VAL	ALA	ASP	GLY	U
	VAL	ALA	ASP	GLY	C
	VAL	ALA	GLU	GLY	A
	VAL	ALA	GLU	GLY	G

The names of the twenty amino acids and their abbreviations are:

ALA	–	Alanine	LEU	–	Leucine
ARG	–	Arginine	LYS	–	Lysine
ASN	–	Asparagine	MET	–	Methionine
ASP	–	Aspartic acid	PHE	–	Phenylalanine
CYS	–	Cysteine	PRO	–	Proline
GLN	–	Glutamine	SER	–	Serine
GLU	–	Glutamic acid	THR	–	Threonine
GLY	–	Glycine	TRP	–	Tryptophan
HIS	–	Histidine	TYR	–	Tyrosine
ILEU	–	Isoleucine	VAL	–	Valine

The abbreviation STOP shows the three triplets which can terminate the polypeptide chain.

RNA and the Genetic Code

RNA is very similar to DNA. Instead of the sugar deoxy-ribose, it has just plain ribose (hence the name *R*ibo*N*ucleic *A*cid), which has an -OH group whose deoxyribose has an -H one. Three of the four bases (A, G and C) are identical to those in DNA. The fourth, *U*racil (U), is a close relative of *T*hymine (T), since thymine is just uracil with a $-CH_3$ group replacing an -H group. This has little effect on the base-pairing. U can pair with A, just as, in DNA, T pairs with A. RNA might be described as using the same language as DNA but with a different accent. RNA can form a double helix, similar but not quite identical to the DNA double helix. It is also possible to form a hybrid double helix which has one chain of RNA and one of DNA. By and large, long RNA double helices are rare, RNA molecules being typically single-stranded, though often folded back on themselves to form short stretches of double helix.

In modern organisms we find RNA used for three pur-poses. For a few small viruses, such as the polio virus, it is used instead of DNA as the genetic material. Some viruses employ single-stranded RNA; a few use it double-stranded. RNA is also used for structural purposes. The ribosomes, the complex assembly of macromolecules which are the actual site of protein synthesis, are made of several structural RNA molecules, assisted by several tens of distinct protein mole-cules. The molecules which act as the interface between the amino acid and the triplet of bases associated with it are also made of RNA. This family of RNA molecules, called tRNA (for *t*ransfer RNA), are used to carry each amino acid to a ribosome, where it will be added to a growing polypeptide chain which will, when complete, become a folded protein.

The third and perhaps the most important use the cell

makes of RNA is as messenger RNA. The cell does not use the DNA itself for everyday work but instead keeps it as the file copy. For working purposes it makes many RNA copies of selected parts of the DNA. It is these tapes of messenger RNA which direct the process of protein synthesis on the ribosomes, using the genetic code outlined in the Appendix.

In any detailed discussion of the origin of life the properties of tRNA molecules loom large, since there is a strong suspicion that they, or a simplified version of them, first arose, if not at the actual beginning of self-replicating systems, at least not long after. Single-stranded nucleic acid molecules, and RNA in particular, will often fold up on themselves, turning back to make short lengths of double helix where the base-sequence permits. tRNA molecules are an excellent example of this. The backbone does not dangle all over the place, but has folded into a relatively compact and rather intricate structure. This exposes at one point a set of three bases (called the *anticodon*) which pairs with the appropriate three bases (called the *codon*) on the messenger RNA. The tRNA acts as an adaptor, with an amino acid at one end and the anticodon at the other, since there is no mechanism by which an amino acid can recognize the codon (the appropriate base-triplet on the messenger RNA) in a direct manner. The specificity of the present-day genetic code is thus embodied in the set of tRNA molecules, at least one type (and usually more) for each amino acid, and also in the set of twenty enzymes (one for each amino acid) which joins each amino acid to the appropriate tRNA molecules. The information for the production of all these essential components for protein synthesis (and many more) is nowadays encoded in the genes, in the appropriate stretches of DNA.

FURTHER READING

Goldsmith, Donald, Editor, *The Quest for Extraterrestrial Life. A Book of Readings.* Mill Valley, California: University Science Books, 1980.

This is a collection of papers (with a commentary by the editor) starting with one by Giordano Bruno, published in 1584. Most of the papers are fairly recent. It includes the original one Orgel and I wrote on Directed Panspermia. Some of them are rather technical, but the general reader should be able to capture the flavor of most of them.

Jastrow, Robert, and Malcolm M. Thompson, *Astronomy: Fundamentals and Frontiers.* New York: John Wiley and Sons, Inc., Second Edition, 1972.

An excellent text for the educated reader, since it is specifically aimed at liberal arts students. Even though it uses no mathematics it gets the ideas over at a fairly advanced level. The weakest part is the section toward the end on viruses.

Miller, Stanley L., and Leslie E. Orgel, *The Origins of Life on Earth*. Englewood Cliffs, New Jersey: Prentice-Hall, Inc., 1974.
> An advanced treatment, suitable for someone with a science degree of some sort.

Orgel, Leslie E., *The Origins of Life*. New York: John Wiley and Sons, 1973.
> This very readable book is rather more difficult than the present text, being written with high school students with advanced scientific training in mind, but it can still be understood by the general reader.

Stryer, Lubert, *Biochemistry*. San Francisco: W. H. Freeman & Company, Second Edition, 1981.

Watson, James D., *Molecular Biology of the Gene*. Menlo Park, California: W. A. Benjamin, Inc., Third Edition, 1976.
> These two books are really textbooks for college students but they are both written in a breezy, readable manner and could serve as further reading for anyone wanting to go more deeply into the chemical foundations of modern biology. The difference between biochemistry and molecular genetics need not detain us. As Arthur K. Fritzmann has said, "We're all molecular biologists now."

INDEX